THE MASTER HANDBOOK OF
ACOUSTICS

BY F. ALTON EVEREST

TAB BOOKS Inc.
BLUE RIDGE SUMMIT, PA. 17214

FIRST EDITION

THIRD PRINTING

Printed in the United States of America

Reproduction or publication of the content in any manner, without express permission of the publisher, is prohibited. No liability is assumed with respect to the use of the information herein.

Library of Congress Cataloging in Publication Data

Everest, F. Alton (Frederick Alton), 1909-
    The master handbook of acoustics.

    Includes index.
    1. Sound. 2. Electronics. I. Title.
QC225.15.E93   693.8′34   81-9212
ISBN 0-8306-0008-6   AACR2
ISBN 0-8306-1296-3 (pbk.)

Contents

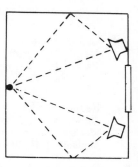

Dedication

To Elva

Foreword

Directly or indirectly, all questions connected with this subject must come for decision to the ear, as the organ of hearing; and from it there can be no appeal. But we are not therefore to infer that all acoustical investigations are conducted with the unassisted ear. When once we have discovered the physical phenomena which constitute the foundation of sound, our explorations are in great measure transferred to another field lying within the dominion of the principles of Mechanics. Important laws are in this way arrived at, to which the sensations of the ear cannot but conform.

Lord Rayleigh in
The Theory of Sound
First Edition 1877
(Also in first American
edition, 1945.)
(courtesy of Dover
Publications Inc.).

Introduction

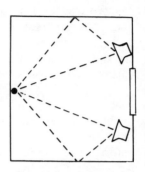

It is a hard fact of life that the intangibility of the acoustics link in the audio chain tends to obscure its vital importance. Hands on experience with microphones, amplifiers and loudspeakers lends a feeling of familiarity and comfort almost entirely lacking in our attitude toward the acoustical environment in which the microphone and loudspeaker function. There seems to be a great gulf in our understanding between the experience of listening to the music of Beethoven, Berlioz or the Bee Gees and fluctuating molecular densities in the air, and even less to the tickling of hair cells in the organ of Corti of the inner ear.

Yet acoustics, the science of sound, has two natures, physical and psychophysical. Sound as a disturbance in air is physical, sound as perceived by the ear is psychophysical. The old conundrum, "If a tree falls in the forest with no ear to hear it, is sound produced?", precisely distinguishes between sound as a stimulus and sound as a sensation. The study of sound as a physical stimulus alone is commonly viewed as a rather sterile subject of only academic interest. As a sensation, however, sound impinges on human experience and becomes of intense interest to us.

This book treats both the physical and psychophysical aspects of sound because the two are so inextricably interrelated. Whether the end product is a recording, a radio or television program or a live performance, the human ear-brain mechanism is intimately involved. In the electronic media, room acoustics is involved twice, once in the pickup and recording in the studio and again in reproduction in the home or classroom. Human ears listen and evaluate at both ends of the process.

By considering the dual aspect of sound, this book consistently considers the response of the human ear to various acoustical effects. Thus echoes affect intelligibility, room resonances affect quality of perceived sound and the combination of two or more acoustical signals can distort the sound heard, and so on. Some chapters treat sound phenomena as physical effects, but as a background for other chapters concentrating on one of the most amazing mechanisms in all nature, the human ear and the acoustic cortex of the human brain.

Of the many ways sound may be put to use, this book concentrates on sound as a medium of communication. The sound might be canned on film, tape or laser disk in analog or digital form, or it might be live. The information to be communicated may be for entertainment, instruction or just carrying on the multifaceted business of daily living. Whether professionals or amateurs are involved, whatever the medium or the purpose, the very word communication implies a junction or a joining, a sender and receiver, one or more persons at each end. If we recognize this personal aspect of the process and the existence of possible perceptual gaps and gulfs, we are in a better position to achieve true and accurate communication.

I am deeply in debt to the scores of authors of technical and scientific papers referred to throughout this book. Surely, it is a great privilege to stand on such stalwart shoulders as the field of acoustics is reviewed. I am grateful to Martin Gallay of Gallay Communications, Inc., for permission to use material first appearing in Recording Engineer/Producer Magazine, principally in Chapters 5, 7, 10, and 12. As for the accuracy of the text, I assume full responsibility. My confidence in doing so, however, is strengthened considerably because of the checking of several key chapters by Robert S. Gales and Robert W. Young, former associates in underwater sound research who are now acoustical consultants in San Diego.

First, Some
Fundamentals of Sound

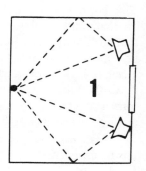

It may be considered ironic to go back to 1866 for our introduction to some fundamentals of sound, but why not if by so doing we can avoid some of the ho-hum mathematics and hackneyed, opaque statements associated with most introductions to the science of sound? In 1866 a German by the name of Kundt published a paper describing an interesting experiment which has become firmly embedded in Physics 100 type courses around the world. Kundt's tube is nothing more than a glass or plastic tube with a source of sound at one end and a plug at the other as shown in Fig. 1-1. A dust-like powder is placed inside the tube. In the olden days lycopodium powder was found both suitable and readily available, but refined talc or finely ground cork will also perform satisfactorily.

The whole point of Kundt's tube (K-tube) is to make visible the effects of air-molecule movements which are invisible to the unaided eye. Sound requires an elastic medium for its propagation and air is elastic. If air molecules are moved out of position by an outside force, they tend to spring back to their original position if the force is removed. A vibrating diaphragm in the sound source compresses the adjacent air molecules when it moves outward and thins them out when it is retracted. If the diaphragm is vibrated continuously, this molecular compression and thinning out (rarefaction) is passed on to adjoining molecules. Although each molecule vibrates back and forth always close to home, it impacts others nearby and the energy transmitted via molecular collision constitutes the propagation of sound.

Figure 1-1 shows the piles of dust powder in the K-tube when the sound source is energized by an amplifier driven by a sine wave from an oscillator. By varying the oscillator frequency we note that the dust powder collects in orderly piles at certain frequencies. The powder concentrates where the molecular movement is minimum in the tube. With a little experimentation we discover that holding the oscillator frequency constant, the same effect can be achieved by varying the position of the plunger-plug. By cutting a couple of corners we conclude that the sound wave propagated down the tube and the wave reflected from the plug combine to form a standing wave, a sort of resonance, when the powder piles are highest and most agitated at the fringes.

When oscillator frequency or plunger position is changed for maximum effect on the powder, we sense that the sound in the K-tube and the distance between vibrating diaphragm and plug have arrived at some sort of important relationship. This relationship can be explained by reference to Fig. 1-1 again. The piles of powder occur at points of zero or minimum sound pressure (nodes), points A, C, E, etc. Between A and C zeros there is a peak (anti-node) of positive pressure at B where air molecules are compressed and crowded together (compression). A corresponding negative peak occurs at D where air molecules are thinned out (rarefaction). These dynamic positive and negative sound pressures fluctuate about the "zero" pressure which turns out to be, not a zero but the prevailing atmospheric pressure. Let one thing be clear, however: the fluctuations are highly exaggerated in Fig. 1-1. The pressure scale has a squiggle in it to indicate it has been shortened. In true scale, the sound pressure fluctuations on this sketch of, say, one unit are to be compared to something like a million units of atmospheric pressure. Even though minute, the sound pressure fluctuations of even weak sounds are readily detected by the ear and measured by meters.

Sound itself is a very objective fact of life familiar to all. The part which the air medium plays in propagating sound, however, is very abstract because it involves invisible air molecules. Smoke particles, which can be seen with the aid of a low-power, darkfield microscope, are small enough to act much like air particles as we look into the propagation of sound. By injecting some smoke into our K-tube we can see how air molecules act as they are set in vibration by passing sound waves. The illuminated smoke particles, normally appearing as tiny white dots, appear as lines under the influence of sound as seen in Fig. 1-2. The length of each white

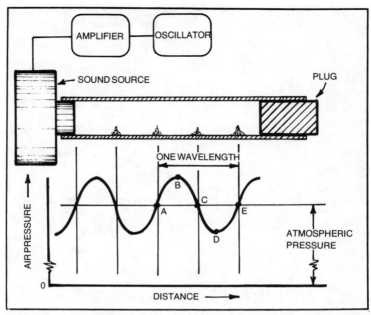

Fig. 1-1. In the Kundt's tube the wave traveling to the right from the source and the wave reflected from the end plug come into coincidence at certain frequencies. This results in a standing wave which has nodes of zero pressure and anti-nodes of high pressure. Bits of cork in the tube collect at the nodes where air particle motion is a minimum. By measuring the distance between nodes and knowing the frequency of the sound, the velocity of sound may be determined.

line is the peak-to-peak amplitude of vibration of that particular smoke particle. For a very loud sound this would be only about 4 thousandths of an inch (0.1 mm).

Energy is required to arrange the powder particles in the K-tube of Fig. 1-1 into neat piles. The reality of sound energy is dramatically illustrated in the ancient experiment shown in Figs. 1-3 A&B. At the focal point of the parabolic reflector is an ultrasonic whistle, called a Galton whistle after its inventor, blown by compressed air and capable of producing intense sound. Although out of the range of the human ear, any dogs around would be greatly agitated at such an uncomfortably loud sound. From the parabolic reflector come plane waves which are reflected from the glass surface establishing a standing wave condition. Using narrow tweezers to avoid disturbing the sound field, thin slices of cork may be inserted at nodal points, resulting in the stack of cork chips shown which have no visible means of support. The force of the sound wave is supporting the cork chips through the violent up and down dance of the air molecules.

11

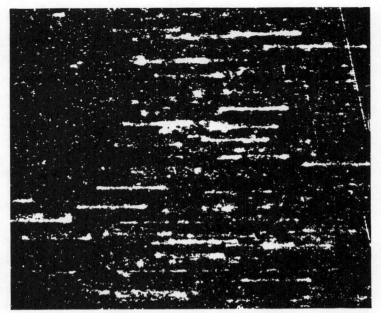

Fig. 1-2. Smoke particles in a Kundt's tube viewed with a microscope appear as white spots with the sound turned off and as white lines with the sound turned on as they follow the same motion as the air particles when sound is propagated (from Alexander Wood's *Acoustics*, Reference 1).

FREQUENCY

In Fig. 1-1 the diaphragm of the sound source is driven back and forth by the amplifier. Each round trip of the diaphragm, from neutral position to maximum deflection in one direction, back to neutral and on to a maximum position in the opposite direction and back to neutral, constitutes one cycle of diaphragm travel and one cycle of sound is produced (A to E). The number of these cycles in one second is called the frequency of the sound. In honor of an early experimenter with electromagnetic radiation, one cycle per second is called one hertz (symbol, Hz). One thousand cycles per second is called one kilohertz (kHz), and so on. Today we have electronic frequency counters which are capable of toting up every positive loop in a given length of time (commonly 1 or 0.1 second) and register the total on a readout display.

WAVELENGTH

The wavelength of a sound is the distance a sound wave travels in one cycle, or 1 Hz. This is indicated in Fig. 1-1 as the distance between alternate zero (or neutral point) crossings, such

12

as A and E. It is the same distance between adjacent positive peaks or adjacent negative peaks or the distance between any two corresponding points of a given cycle. It is quite evident in studying Fig. 1-1 that frequency and wavelength are related. This relationship also involves the speed of sound:

$$\lambda = \frac{c}{f} \tag{1-1}$$

where, λ = wavelength of sound in feet (or meters),
  c = speed of sound, feet per second (or meters per second) and
  f = frequency, hertz or cycles per second.

The K-tube of Fig. 1-1 now becomes a research instrument. How fast does sound travel? is the question to be answered. The K-tube is fired up and the distance between alternate piles of powder is scaled off (e.g., A to E) and found to be about 2¼ inches. Without an electronic frequency meter an estimate of the frequency is read off the oscillator dial as about 6,000 Hz. From the above equation the speed of sound is found to be (6,000) (2.25/12) = 1125 feet per second. Such precision shouldn't happen in such a crude device! The accepted value of speed of sound at air temperature 68°F (20°C) is actually 1127 feet per second.

It is interesting that for audible sounds within the range of our ears the wavelengths also fall within a familiar range. Taking the speed of sound as 1127 ft. per sec., wavelengths of typical audible frequencies are as follows:

Frequency	Wavelength	
	inches	centimeters
20 Hz	676.2	1,717.5
1 kHz	13.5	34.4
8 kHz	1.7	4.3
20 kHz	0.68	1.7

Later we shall see how the fact that bass sounds having wavelengths comparable to room dimensions create some very special problems.

THE SIMPLE SINUSOID

The sound source diaphragm moving in and out produces sound pressure fluctuations of the shape shown in Fig. 1-1. This is

13

called a sine wave. The sine form is directly related to what is called simple harmonic motion. A weight vibrating on a spring displays such motion. The piston in an automobile engine is connected to the crankshaft by a connecting rod. The crankshaft going around and around and the piston up and down beautifully illuminate the inherent relationship between rotary motion and linear simple harmonic motion. The piston position (neglecting connecting rod angle) plotted against time produces a sine wave. It is a very basic type of mechanical motion and it yields an equally basic waveshape in sound and electronics.

Fig. 1-3A. The Galton whistle blown by compressed air demonstrates the force of sound waves as slices of cork are levitated. A standing wave system is set up between the parabolic reflector with the ultrasonic whistle at the focal point and the plate glass surface (courtesy of Moody Institute of Science).

Fig. 1-3B. Close-up view of Fig. 1-3A (courtesy of Moody Institute of Science).

COMPLEX WAVES

Speech and music waveshapes depart radically from the simple sine form. A very interesting fact, however, is that no matter how complex the wave, if it is periodic, it can be reduced to sine components. The obverse of this is that any complex wave can, theoretically at least, be synthesized from sine waves of different frequencies, different amplitudes, and different time relationships (phase). A friend of Napolean named Fourier initiated thinking in this surprising direction. This can be viewed as either a simplification or complication of the situation; certainly it is a great simplification in regard to concept, but sometimes complex in application to specific speech or musical sounds. As we are interested primarily in basic concept, let us see how even a very complex wave can be reduced to simple sinusoidal components.

HARMONICS

A simple sine wave of a given amplitude and frequency f_1 is shown in Fig. 1-4A. In Fig. 1-4B is another sine wave of half the amplitude and twice the frequency (f_2). Combining A and B the waveshape of Fig. 1-4C is obtained. In Fig. 1-4D another sine wave of half the amplitude of A and three times its frequency (f_3) is shown. Adding this to the $f_1 + f_2$ wave of C, Fig. 1-4E is obtained. The simple sine wave of Fig. 1-4A has been progressively distorted as other sine waves have been added to it. Whether these are acoustic waves or electronic signals the process can be reversed. The distorted wave of Fig. 1-4E can be disassembled, as it were, to the simple f_1, f_2, and f_3 sine components by either acoustical or electronic filters. For example, passing the wave of Fig. 1-4E through a filter passing only f_1 and rejecting f_2 and f_3, the original f_1 sine wave emerges in pristine condition.

· Applying names, the sine wave of lowest frequency (f_1) of Fig. 1-4A is called the *fundamental*, the one of twice the frequency (f_2) of Fig. 1-4B is called the *second harmonic* and the one of three times the frequency (f_3) of Fig. 1-4D is the *third harmonic*. The fourth harmonic, the fifth harmonic, etc., are, of course, four and five times the frequency of the fundamental, and so on. Curiously, even though the waveshape is dramatically changed by shifting time relationships, the ear is relatively insensitive to such time changes although quite sensitive to harmonic amplitude content.

PHASE

All three components, f_1, f_2, and f_3, start from zero together. This is called an in-phase condition. In some cases the time relationships between harmonics or between harmonics and the fundamental are quite different than this. Remember how one revolution of the crankshaft of the automobile engine (360°) was equated with one cycle of simple harmonic motion of the piston? The up and down position of the piston spread out in time traces a sine wave such as that in Fig. 1-5. One complete sine wave cycle represents 360° of rotation. If another sine wave of identical frequency is delayed 90°, its time relationship to the first one is a quarter wave late. A half wave delay would be 180°, etc. For the 360° delay the wave at the bottom of Fig. 1-5 falls in step with the top one, reaching positive peaks and negative peaks simultaneously, the in-phase condition.

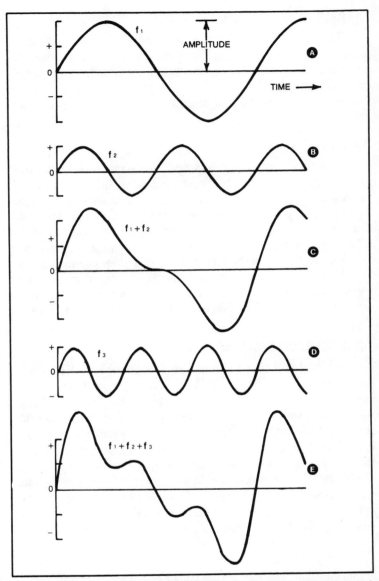

Fig. 1-4. A study in the combination of sine waves. (A) the fundamental of frequency f_1. (B) A second harmonic of frequency $f_2 = 2f_1$ and of half the amplitude of f_1. (C) The sum of f_1 and f_2 obtained by adding ordinates point by point. (D) A third harmonic of frequency $f_3 = 3f_1$ and of half the amplitude of f_1. (E) The waveshape resulting from the addition of f_1, f_2, and f_3. All three components are "in phase", i.e., they all start from zero at the same instant. Shifting the time relationship of f_2 and/or f_3 with respect to f_1 would alter the shape of (E) materially.

PARTIALS

The musician is inclined to use the term *partials* instead of harmonics, and it is well that a distinction is made because the partials of many musical instruments are not harmonically related to the fundamental. Partials may or may not be exact multiples and richness of tone may be imparted by such deviations from true harmonic relationship.

OCTAVES

Audio and electronics engineers and acousticians have frequent occasion to use the linear integral multiple concept of harmonics, closely allied as it is to the physical aspect of sound. The musician often refers to the octave, a logarithmic concept, firmly embedded in musical scales and terminology because of its relationship to the ear's characteristics. Audio people, of course, are also involved with the human ear, hence their common use of logarithmic scales for frequency, logarithmic measuring units, and various devices based on octaves which will be more fully developed later. Harmonics and octaves are compared in Fig. 1-6.

THE CONCEPT OF SPECTRUM

In Chapter 2 the commonly accepted scope of the audible spectrum from 20 Hz to 20 kHz will be related to specific characteristics of the human ear. Here, in the context of sine waves, harmonics, etc., we need to establish the spectrum concept. The visible spectrum of light has its counterpart in sound in the audible spectrum, the range of frequencies which fall within the perception limits of the human ear. We cannot see the far ultraviolet light because the frequency of its electromagnetic energy is too high for the eye to perceive. Nor can we see the far infrared light because its frequency is too low. There are likewise sounds of too low and too high frequency for the ear to hear, infrasound and ultrasound.

In Fig. 1-7 are several waveforms which typify the infinite number of different waveforms commonly encountered in audio. These waveforms have been photographed directly from a cathode ray oscilloscope. To the right of each photograph is the spectrum of that particular signal. This spectrum tells how the energy of that signal is distributed in frequency. In all but the bottom signal of Fig. 1-7 the audible range of the spectrum was searched with a wave analyzer having a very sharp filter with a passband only 5 Hz wide. In this way concentrations of energy were located and measured with an electronic voltmeter.

For the ideal sine wave, all the energy is concentrated at one frequency. This particular sine wave from this particular signal generator is not what could be called a pure sine wave. No oscillator is perfect, all have some harmonic content, but scanning the spectrum of this sine wave the harmonics measured were too low to show on the graph scale of Fig. 1-7.

The triangular wave has a major fundamental component of 10 units magnitude. The wave analyzer detected a significant 2nd harmonic component at f_2, twice the frequency of the fundamental

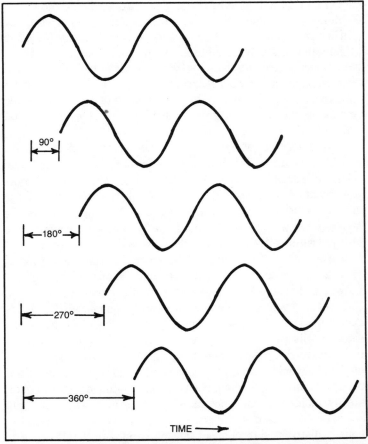

Fig. 1-5. Illustrating phase relationships. There is a basic relationship between the back and forth motion of a loudspeaker diaphragm, for example, and rotary motion. Rotation of 360° is analagous to one complete sine cycle.

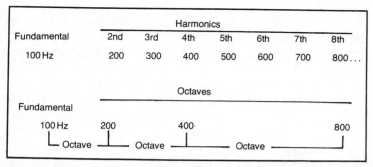

Fig. 1-6. Comparison of harmonics and octaves. Harmonics are linearly related, octaves logarithmically.

having a magnitude of 0.21 units. The third harmonic showed an amplitude of 1.13 units, the 4th of 0.13 units, etc. The 7th harmonic still had an amplitude of 0.19 units and the 14th harmonic (about 15 kHz in this case) an amplitude of only 0.03 units, but easily detectable. So we see that this triangular wave has both odd and even components of modest amplitude down through the audible spectrum. Knowing the amplitude and phases of each of these and combining them, the original triangular wave shape could be synthesized.

A comparable analysis reveals the spectrum of the square wave shown in Fig. 1-7. It has harmonics of far greater amplitude than the triangular wave with a distinct tendency toward more prominent odd harmonics than even. The 3rd harmonic showed an amplitude 34% of the fundamental! The 15th harmonic of the square wave was still 0.52 units! If the synthesis of a square wave stopped with the 15th harmonic, the wave of Fig. 1-8C results.

A glance at the spectra of sine, triangular, and square waves reveals energy concentrated at harmonic frequencies, but nothing between. These are all so-called periodic waves which repeat themselves cycle after cycle. The fourth example in Fig. 1-7 is random noise. The spectrum of this signal cannot be measured satisfactorily with a wave analyzer with a 5 Hz passband because the fluctuations are so great it is impossible to get a decent reading on the electronic voltmeter. Analyzed by a wider passband of fixed bandwidth and with the help of various integrating devices to get a steady indication, the spectral shape shown is obtained. This spectrum tells us that the energy of the random noise signal is equally distributed throughout the spectrum until the drooping at high frequencies indicates that the upper frequency limit of the random noise generator has been reached.

There is little visual similarity between the sine and the random noise signals as revealed by the cathode ray oscilloscope, yet there is a hidden relationship. Even random noise can be considered as made up of sine wave components constantly shifting in frequency, amplitude, and phase. Passing random noise through a narrow filter and observing the filter output on a cathode ray oscilloscope, a restless, sine-like wave is seen which constantly shifts in amplitude. Theoretically, an infinitely narrow filter would sift out a pure, but nervous, sine wave.

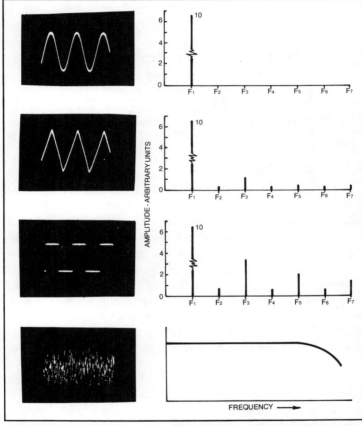

Fig. 1-7. The spectral energy of a pure sinusoid is all concentrated at a single frequency. The triangular and square waves each have a prominent fundamental and numerous harmonics at integral multiples of the fundamental frequency. Random noise (white noise) has energy distributed uniformly throughout the spectrum up to some point at which energy begins to fall off due to generator limitations. Random noise may be considered a mixture of sine waves of continuously shifting frequency, amplitude, and phase.

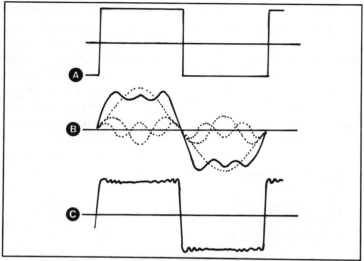

Fig. 1-8. In synthesizing the square wave of (A), including only the fundamental and two harmonics yields (B). Including fifteen components yields (C). It would take many more than fifteen harmonics to smooth the ripples and produce the square corners of (A).

WHITE AND PINK NOISE

Random noise is random noise but the literature is filled with references to *white noise* and *pink noise*. What is the difference? These two colorful terms have grown out of the fact that there are two types of spectrum analyzers in common use. One, typified by the 5 Hz analyzer mentioned in the previous section, is the constant bandwidth analyzer. No matter whether the analyzer is scanning the low frequencies or the high, the passband remains a fixed number of hertz wide. When random noise is scanned by such an analyzer, its uniform distribution of energy yields the flat curve of Fig. 1-7. The random noise we have been referring to up to this point is called *white noise* because its counterpart in light is white light.

The second type of analyzer in common use has a bandwidth which increases with frequency. Examples of this are 1/3 octave, 1 octave, or other analyzers having passband widths a constant percentage of the band center frequency. One octave centered at 125 Hz (89-177 Hz) spans 88 Hz and has a bandwidth of 70.7% (88/125 x 100). An octave centered on 8 kHz (5,656 - 11,312 Hz) spans 5,656 Hz and also has a bandwidth of 70.7% (5,656/8,000 x 100). If the noise signal of Fig. 1-7 were analyzed with a filter of this type, the spectrum graph would slope upwards with a slope of 3 dB/octave. Because noise signals are so desirable in acoustical measurements, and to avoid this deceptive upward sloping spectrum, random noise is commonly filtered to give a compensating 3 dB/octave downward slope. Such a random noise signal with a 3 dB/octave downward slope is called pink noise because the low frequency end (the red end of the visible spectrum) is emphasized.

Converting white noise to pink noise is accomplished by sending it through a so-called pink noise filter. Such a filter is shown in Fig. 1-9 and is nothing more than four resistors and four capacitors connected as shown. Figure 1-10 shows rough measurements made in analyzing white noise (A) with an octave analyzer. As previously stated, when the spectrum of white noise is analyzed with octave filters, it rises with frequency with a slope of 3 dB/octave. The reason for this is that there is a fixed amount of energy per hertz and the higher the octave the more hertz embraced and hence the higher the energy level.

Figure 1-10(B) repeats the octave analysis of white noise as it is passed through the filter of Fig. 1-9. In other words, the filter characteristic falls off with frequency at 3 dB/octave, compensating for the rising white noise spectrum of Fig. 1-10(A) resulting in the flat pink noise characteristic of Fig. 1-10(B). This pink noise, being flat throughout the audible spectrum (while using constant percentage filters), is convenient to detect variations from a flat

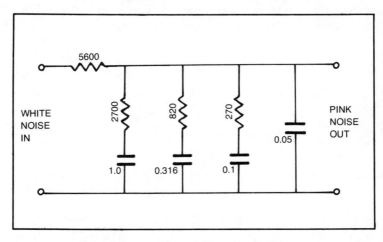

Fig.1-9. A simple filter for changing white noise to pink noise. It changes random noise of constant energy per Hertz to pink noise of constant energy per octave. Pink noise is useful in acoustical measurements utilizing analyzers having passbands of width a constant percentage of the center frequency (courtesy of General Radio Com pany, Reference 2).

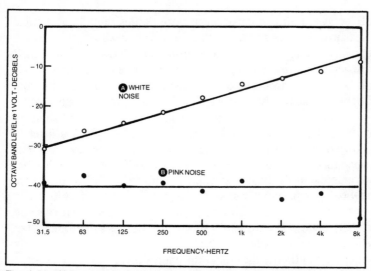

Fig. 1-10. White noise, having uniform distribution of energy with frequency, shows the spectrum of (A) when analyzed with octave filters. The spectrum rises at 3 dB per octave because the octave filters have widths of ever increasing numbers of Hertz. Pink noise shows a flat spectrum when analyzed with octave filters. This flatness makes pink noise convenient for certain acoustical measurements.

response of a room, an amplifier, a loudspeaker, or other acoustical or electronic component. The real time analyzer employing 1/3 octave bands, to be discussed more fully in a later chapter, is almost always used with pink noise as the excitation signal.

Hearing— That Marvelous Second Sense

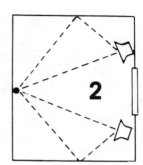

2

No one deeply involved in audio would be inclined to minimize the importance of sight, the so-called first sense, but a very special interest in the sense of hearing is understandable for obvious reasons. We are told that 83% of what we learn comes to us through the eye gate and only 11% through hearing. But when it comes to recalling what is learned at a later time, a combination of sight and sound is far superior to either alone. The phenomenal success of audiovisual techniques in education bears this out. The senses were undoubtedly intended to serve our needs, each in its own specific way and elevating one in importance over another is futile. Each is a window of our body-house opening on to a different aspect of our environment, each bringing different input for analysis and appropriate action.

Audio workers can be forgiven for a special interest in the human ear-brain mechanism and the amazing sense of hearing. Knowledge of the inner workings of our hearing impinges directly on our day to day work, whether in electronics, high fidelity, music, recording, radio and television broadcasting, sound reinforcement, architectural acoustics, or what have you. For that matter, it also directly impinges on those of all trades, professions, and activities in which human speech plays a part and that takes in just about everybody. There is an inevitable acoustical-aural link in the communication chain whenever sound is involved.

SENSITIVITY OF THE EAR

The delicate and sensitive nature of our hearing can be dramatically underscored by a little experiment. Most readers will not have the privilege of participating in this experiment personally, but come with us vicariously! The bulky door of the anechoic chamber is slowly opened revealing extremely thick walls and three foot wedges of glass fiber, points inward, lining all walls, ceiling, and what could be called the floor except that one walks on an open steel grillwork above it. A chair is brought in and you sit down on it. This experiment takes time and, as a result of prior briefing, you lean back, patiently counting the glass fiber wedges to pass the time. It is really eerie in here. The sea of sound and noises of life and activity in which we are normally immersed and of which we are scarcely conscious is now most conspicuous by its absence. The silence presses down on you in the tomb-like silence. Ten minutes. A half hour. New sounds are discovered, sounds that come from within your own body. First, the loud pounding of your heart, still recovering from the novelty of the situation. An hour goes by. The blood coursing through the vessels becomes audible. At last, if your ears are keen, your patience is rewarded by a strange hissing sound between the "ker-bumps" of the heart and the slushing blood. What is it? It is the sound of air particles pounding against your eardrums. The eardrum motion resulting from this hissing sound is unbelievably small only 1/100th of a millionth of a centimeter, or 1/10th diameter of a hydrogen molecule!

The human ear cannot detect softer sounds than the rain of air particles on the eardrum. This is the threshold of hearing. There would be no reason to have ears more sensitive, because any sounds of lower level than that produced by air particle activity would be drowned by the air particle noise. This means that the ultimate sensitivity of our hearing just matches the softest sounds possible in an air medium. Accident? Adaptation? Design? Draw your own conclusions.

At the other extreme our ears can respond to the roar of a cannon, the noise of a rocket blastoff, or a jet aircraft under full power. Special protective features of the ear protect the sensitive mechanism from damage from all but the most intense noises.

A PRIMER OF EAR ANATOMY

The arrangement of the principal parts of the ear are shown in Fig. 2-1. The ear canal is called the *external meatus*. At the inner end of this canal is the eardrum. This is the only part of the ear to which we have direct access and, to the horror and consternation of ear specialists, is the part into which children are prone to stuff a variety of foreign objects and from which some adults remove wax with paper clips or matches.

The third section (Fig. 2-1), the inner ear, is the sensory transducer. Here, imbedded in solid skull bone, the mechanical vibrations are translated into nerve impulses which are sent to the acoustic cortex of the brain. Still shrouded in mystery after decades of intense research, the inner ear is slowly, if reluctantly, yielding the secret of its extremely complex operation.

Between the eardrum and the inner ear is an air-filled cavity housing the mechanical linkage between the eardrum and the inner ear. Three tiny bones, the ossicles, provide this linkage. Motions of the eardrum resulting from external sounds are transmitted through these tiny bones to the fluid-filled inner ear. The airborne sound falling on the external ear is changed to mechanical motion of the ossicles which, in turn, set up corresponding sound waves in the fluid of the inner ear. Some people are surprised to learn that sound travels through solids and liquids, but remember how the neighbor's hi-fi sounds readily penetrate a common wall and how porpoises talk to each other by sound carried by sea water. The ear exemplifies sound traveling through air, liquid, and solid bone (bone conduction) as well as by mechanical linkage of tiny bones.

The Ear Canal

Not much attention will be given to the outer ear, called the *pinna*, which, among other functions, gathers sound and directs it into the ear canal. Cupping one's hand to the ear

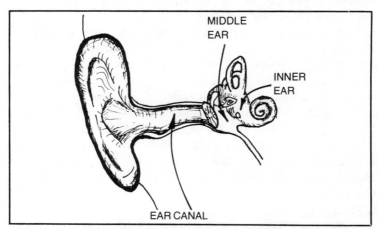

Fig. 2-1. The three principal parts of the human ear, the auditory canal, the middle ear, and the inner ear.

increases the loudness of sound a noticeable amount. The external ear acts in a similar way.

The ear canal also increases the loudness of sounds we hear. In Fig. 2-2 this ear canal with a diameter of about 0.7cm and about 3 cm long is idealized by straightening it out and making it uniform in diameter. Acoustically, this can be taken as quite a close representation. It is a pipe-like duct, closed at the inner end by the eardrum. Organ pipes were studied intensely by early investigators when the science of acoustics was in its infancy. The acoustical similarity of this ear canal to an organ pipe was not lost on early workers in the field. A resonance effect of the ear canal is expected, increasing sound pressure at the eardrum at certain frequencies. The maximum effect is near the frequency at which the 3cm pipe is one quarter wavelength. This turns out to be about 3,000 hertz. Figure 2-3(A) shows the results of measurements at different frequencies which show the increase in sound pressure at the eardrum over that at the opening of the ear canal. We note the peak is not far from the rough 3,000 hertz predicted from the 3cm pipe length, the measurements of which are somewhat indefinite because of the uncertainty as to where the ear canal ends and the pinna begins. Organ builders are also aware of an "end effect" which changes the effective length of a pipe. Pipe resonance, we conclude, amplifies the sound pressure falling on the outer ear about 3 fold (10 dB) by the time it strikes the eardrum, peaking in the 2 - 4 kHz region.

There is a lot more involved than pipe resonance. Wiener and Ross have found (for sound arriving face-on) that diffraction around the head results in a further amplification effect.[3,4] Diffraction is the deflection of sound into a shadow zone and the human head is dense enough (no pun intended) to create sound shadows. Strangely enough, the head disturbs a diffuse (thoroughly mixed) sound field in such a way as to result in a sort of diffraction resonance which results in further amplification of sound entering the auditory canal. Line B in Fig. 2-3 demonstrates that line A is only part of the acoustical amplification effect. The overall amplification of sound pressure is of the order of 10-fold (20 dB), still peaking in the 2 - 4 kHz region. It is another marvel of nature that this peaking effect occurs in the same frequency region in which speech energy is concentrated.

The Middle Ear

After spending a number of years in underwater research I am greatly impressed with a number of things, among them the efficiency with which sound is propagated thousands of miles in the sea, and the number and variety of sound producing organisms living in the oceans (and the difficulty of obtaining significant measurements at sea with the green faced technical staff draped over the lee rail hoping for the research vessel to sink and end it all). But one thing stands out above all: the difficulty of transmitting sound energy from a tenuous medium such as air into a dense medium like water. Without some very special equipment, sound originating in air is bounced off a water surface like light off a mirror. It boils down to a matter of matching impedances, and in this case the impedance ratio is something like 4,000 to 1. Consider how satisfactory it would be to drive a 1 ohm voice coil of a loudspeaker with an amplifier having an output impedance of 4,000 ohms! Surely, not much power would be transferred.

The identical problem confronted the Designer of the middle ear. The sound comes in by vibrations of a rather flimsy diaphragm and the object is to get the feeble energy represented by its vibratory motion transferred with maximum efficiency to the water-like fluid of the inner ear. The two-fold solution is suggested in Fig. 2-4. The three ossicles (hammer, anvil, and stirrup) form a mechanical linkage between the eardrum and the oval window which is in intimate contact with the fluid of the inner ear. The first

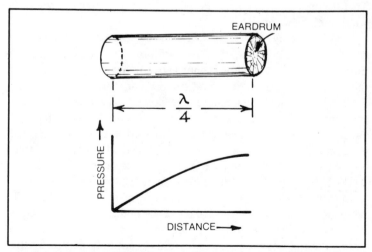

Fig. 2-2. An idealization of the auditory canal to illustrate the acoustical similarity to a pipe closed at one end by the eardrum. Such a pipe is acoustically resonant when the exciting sound is of a frequency to make it a quarter wavelength long. Under these conditions the sound pressure at the eardrum is amplified acoustically.

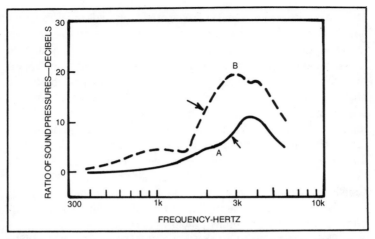

Fig. 2-3. Acoustical amplification of sound pressure at the eardrum. (A) Peak of pressure due to pipe resonance of the auditory canal. (B) Sound pressure at eardrum including effect of (A) plus an added effect of diffusion amplification resulting from the head disturbing a diffuse sound field. An overall 10 fold (20 dB) increase of pressure occurs where speech energy is concentrated (after Wiener and Wiener & Ross, References 3 and 4).

of the three bones, the hammer, is fastened to the eardrum. The third, the stirrup, is actually a part of the oval window. There is a lever action in this linkage with a ratio ranging from 1.3:1 to 3:1. That is, the eardrum motion is reduced by this amount at the oval window of the inner ear.

This is only part of this fascinating mechanical impedance matching device. The area of the eardrum is about 80 sq mm and the area of the oval window is only 3 sq mm, hence a given force on the eardrum is reduced in the ratio of 80/3 or about 27 fold. In Fig. 2-4(B) the action of the middle ear is likened to two pistons having area ratios of 27:1 connected by an articulated connecting rod having a lever arm ranging from 1.3:1 to 3:1, making a total mechanical force increase between 35 and 80. The acoustical impedance ratio between air and water being of the order of 4,000:1, the pressure ratio required to match two media would be $\sqrt{4,000}$ or about 63:1 and we note that this falls within the 35 to 80 range obtained from the mechanics of the middle ear illustrated in Fig. 2-4. The problem of matching sound in air to sound in the water-like fluid of the inner ear is thus beautifully solved by the mechanics of the middle ear. The evidence that the impedance matching plus the resonance amplification of Fig. 2-3 really work is that a diaphragm motion comparable to molecular dimensions gives a threshold perception.

A highly schematic sketch of the ear is given in Fig. 2-5. The conical eardrum at the inner end of the auditory canal forms one side of the air-filled middle ear. The middle ear is vented to the upper throat behind the nasal cavity by the *Eustachian tube*. The eardrum operates as an "acoustic suspension" system, acting against the compliance of the trapped air in the middle ear. The Eustachian tube is suitably small and constricted so as not to destroy this compliance. The round window separates the air-filled middle ear from the practically incompressible fluid of the inner ear. The Eustachian tube fulfills a second function in equalizing the air of the middle ear with outside atmospheric pressure so that the delicate membranes of the inner ear can function properly. Whenever we swallow, the Eustachian tubes are opened, equalizing middle ear and hence the inner ear pressure. When an aircraft undergoes rapid changes of altitude (at least those not having

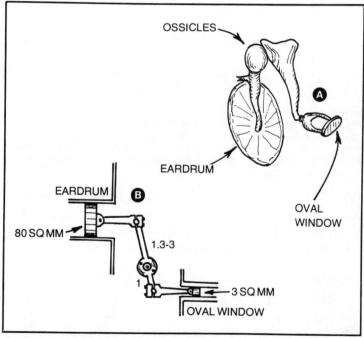

Fig. 2-4. (A) The ossicles (hammer, anvil, and stirrup) of the middle ear which transmit the mechanical vibrations of the eardrum to the oval window of the inner ear. (B) A mechanical analog of the impedance matching function of the middle ear. The differences between eardrum area and oval window area, coupled with the stepdown mechanical linkage, matches the relatively great motions of the eardrum in air to the small motions of the oval window appropriate for working into water.

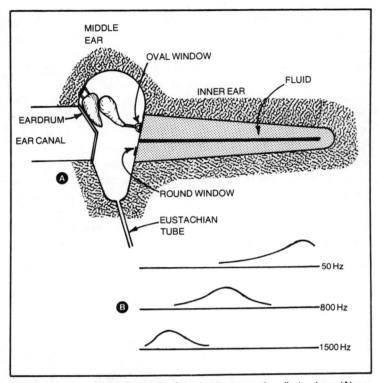

Fig. 2-5. Highly idealized sketch of the human ear primarily to show (A) arrangement of the unrolled, fluid filled cochlea, and (B) how the sound travelling in the fluid of the inner ear causes amplitude peaks to occur at different locations, depending upon the frequency of sound. Hair cells excited by these vibratory peaks send electrical signals to the brain.

pressurized cabins) the occupants may experience momentary deafness or pain until the middle ear pressure is equalized by swallowing. Actually, the Eustachian tube has a third emergency function of drainage if the middle ear becomes infected.

The Inner Ear

That portion of the inner ear housing the auditory equipment (another portion has to do with our sense of balance) is about the size of a pea and is encased in solid skull bone. The *cochlea* is coiled up like a sea shell from which it gets its name. Our descriptive purposes are best served by unrolling this 2-¾ turn coil and stretching it out to its full length, about one inch, as shown in Fig. 2-5. The fluid filled inner ear is divided into a lower and upper part by a pair of membranes. The oval window opens into the upper part and the pressure release round window into the lower part. Vibration of the eardrum sets the ossicles in motion which, in turn, causes a rocking motion of the oval window. This sets up sound waves in the fluid of the inner ear. As the fluid of the inner is nearly incompressible, it would be impossible to set up vibrations in it without the reciprocal action of the round window. When the oval window is driven inward, the round window moves out toward the middle ear. When the cochlea is excited by a 50 Hz sound, a standing wave condition is set up in the cochlea which results in a maximum amplitude near the end away from the oval window as shown in Fig. 2-5B. When the frequency of the sound is changed, the position of this amplitude peak shifts. The peaks shown in Fig. 2-5B are very broad and, of themselves, do not explain the sharpness of frequency discrimination displayed by the ear. Apparently other neural functions downstream have the effect of sharpening these passbands which gives the ear its sharp ability to analyze sounds.

Waves set up in the fluid filled duct of the inner ear act on hairlike nerve terminals which convey signals in the form of neuron discharges to the brain. There are about 24,000 "rods", each with a hair cell from which a dozen or so hairs extend into the cochlear liquid. When sound excites the liquid, membrane and hair cells are stimulated, sending an electrical wave through surrounding tissue. These so-called "microphonic" potentials can be picked up and amplified, reproducing the sound falling on the ear which acts as a veritable biological microphone. These potentials are proportional to sound pressure falling on the ear (linear, that is) over an 80 dB range. The microphonic potentials of the ear of an anesthetized cat have been reproduced for an audience with amazing fidelity.

Apparently, distortions of hair cells trigger the nerve impulses which are carried by the auditory nerve to the brain. While the microphonic signals are analog, the impulses sent to the acoustic cortex are impulses generated by neuron discharges. A single nerve fiber is either firing or not firing (binary!). When it fires it causes an adjoining one to fire and so on. Physiologists liken the process to a burning gunpowder fuse. The rate of travel bears no relationship to how the fuse was lighted. Presumably the loudness of sound is related to the number of nerve fibers excited and the repetition rates of such excitation. When all the nerve fibers (some 3,000 of them) are excited, this is the maximum loudness which can be perceived. The threshold sensitivity would be represented by a single fiber firing. An overall, well accepted theory of how the inner ear and the brain really function is not yet formulated.[5,6,7]

The above is a highly simplified presentation of a very complex mechanism upon which much current research is taking place. Some of the numbers used and theories suggested are not universally accepted. Popularization of a subject such as the ear is an occupation that may be hazardous to the author's health, but any red blooded worker in audio must surely be amazed at the delicate and effective workings of the human ear. I hope also that a new consciousness of, and respect for, this delicate organism will be engendered and that damaging high sound levels will be avoided.

METERS VS. THE EAR

There still remains a great chasm between subjective judgements of sound quality, room acoustics, etc., and objective measurements. During the last decade considerable attention has been focused on the problem. Consider the following descriptive words which are often applied to concert hall acoustics:

warmth	clarity
bassiness	brilliance
definition	resonance
reverberance	balance
fullness of tone	blend
liveness	intimacy
sonority	shimmering

What kind of an instrument measures *warmth* or *brilliance*? How would you devise a test for *definition*? Progress, however, is being made. Take *definition* for instance. German researchers have adopted the term *deutlichkeit* which literally means clearness or distinctness, quite close to *definition*. It is reduced to numbers by measuring the energy in an echogram during the first 50 milliseconds and comparing it to the energy of the entire echogram. This compares the direct sound and early reflections, which are integrated by the ear, to the entire reverberant sound. This relatively straightforward measurement on an impulsive sound from a pistol or pricked balloon holds considerable promise for relating the descriptive term *definition* to an objective measurement. It will be a long time before all of these and a host of other subjective terms can be reduced to objective measurements, but this is the problem in acoustics and psychoacoustics.[8,9,10]

There comes a time at which meter readings must give way to observations by human subjects. Experiments then take on a new, subjective tone. In a loudness investigation panels of listeners are presented with various sounds and each observer is asked to compare the loudness of sound A with the loudness of B or to make judgements in other ways. The data submitted by the jury of listeners are then subjected to statistical analysis and the dependence of a human sensory factor, such as loudness, upon physical measurements of sound intensity is assessed. If the test is conducted properly and sufficient observers are involved, the results are trustworthy. It is in this way that we discover that there is not a linear relationship between sound intensity and loudness, pitch and frequency, nor between timbre and sound quality.[11]

AREA OF AUDIBILITY

The two curves of Fig. 2-6 have been obtained with a group of trained listeners. In this case the listeners face the sound source and judge whether a sound at a given frequency is equal in loudness to another reference tone at 1000 Hz. Each curve is an equal

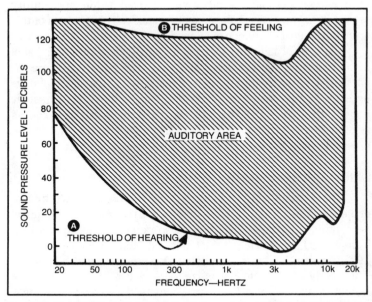

Fig. 2-6. The auditory area of the human ear is bounded by two threshold curves, (A) the threshold of hearing delineating the lowest level sounds the ear can hear, and (B) the threshold of feeling at the upper extreme. All of our auditory experiences must be played in this area bounded by the two thresholds and the frequency range of sounds the ear can perceive.

loudness contour passing through the reference loudness at 1000 Hz. These two curves represent the extremes of our perception of loudness. On the soft end of the loudness scale is our threshold of hearing represented by curve A in Fig. 2-6, which, in the frequency region of most sensitive hearing is near zero dB sound pressure level. At any given frequency the intensity of sound can be increased until a tickling sensation is felt in the ears. This occurs at about a sound pressure level of 120 or 130 dB. Further increase in sound intensity results in an increase of feeling until a sensation of pain is reached. The threshold tickling is a warning that the sound is becoming dangerously loud and that ear damage may be imminent.

Figure 2-6 depicts the two thresholds of the average human ear, the lower threshold below which we cannot hear, and the upper one above which sounds are painful and potentially damaging to the ear mechanism. In between is the area of audibility, an area because it has two dimensions, the vertical dimension of sound pressure level and the horizontal range of frequencies which the ear can perceive. All the sounds which humans experience must be of such frequency and intensity range as to fall within this auditory area. In Chapter 9 we shall see more specifically how much of this area is utilized for common music and speech sounds.

The area of audibility for humans is quite different from that of many animals. The bat specializes in sonar cries which are far above the upper frequency limit of our ears. The hearing of dogs extends higher than ours, hence the usefulness of ultrasonic dogwhistles. Sound in the infrasonic and ultrasonic regions, as related to the hearing of humans, is no less true sound in the physical sense, but it does not result in human perception.

LOUDNESS VS. FREQUENCY

The extremes of the auditory area, Fig. 2-6, marked by two equal loudness contours, is only a partial picture of human perception of loudness. The seminal work on this was done at Bell Telephone Laboratories by Fletcher and Munson and reported in 1933.[12] Since that time refinements have been added by others. The family of equal loudness contours of Fig. 2-7, the work of Robinson and Dadson,[13] has been recommended as an international standard.

The surprising thing about the curves of Fig. 2-7 is that they reveal that perceived loudness varies greatly with frequency and sound pressure level. For example, a sound pressure level of 30 dB yields a loudness level of 30 phons at 1,000 Hz, but it requires a sound pressure level of 58 dB more to sound equally loud at 20 Hz as shown in Fig. 2-8. The curves tend to flatten at the higher sound levels. The 90 phon curve rises only 32 dB between 1,000 Hz and 20 Hz. Putting this effect into perspective, let us consider listening to a symphony orchestra on a home hi fi system. Setting the tone controls at flat position (no high or low frequency boost or cut) a fortissimo passage sounds all right, but a softer passage sounds deficient in bass. This is because the ear is less sensitive to bass notes than midband notes at low levels. There are wiggles in the ear's high frequency response which are somewhat less noticeable. This bass problem of the ear means that the quality of reproduced music depends on the volume control setting. Listening to background music at low levels requires a different frequency response than listening at higher levels.

LOUDNESS CONTROL

The loudness control is a popular approach to this problem, but far from a true solution to it. Think of all the things that affect the volume control setting in a particular situation. The loudspeakers vary in acoustic output for a given input power. The gain of

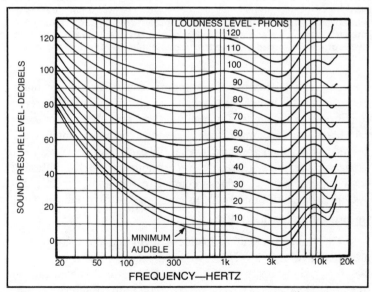

Fig. 2-7. Equal loudness contours as determined by Robinson and Dadson (Reference 13). These contours reveal the lack of sensitivity to bass tones, especially at lower levels. Inverting these curves gives the frequency response of the ear in terms of loudness level.

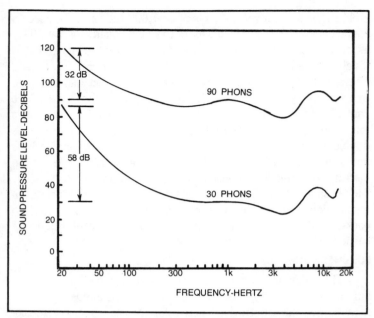

Fig. 2-8. A comparison of the ear's response at 20 Hz compared to that at 1,000 Hz. At a loudness level of 30 phons, the sound pressure level of a 20 Hz tone must be 58 dB higher than that at 1,000 Hz for the two to have the same loudness. At 90 phons loudness level, an increase of only 32 dB is required. The ear's response is somewhat flatter at high loudness levels. Loudness level is only an intermediate step to true subjective loudness as explained in a later section.

preamplifiers, power amplifiers, tuners, and phono pickups differ from brand to brand and circuit to circuit. Listening room conditions vary from dead to highly reverberant. With all of these variables, how can a manufacturer design a loudness control truly geared to the sound pressure level at the ear of listener X with X's particular variables geared to X's equipment and X's listening environment? For a loudness control to function properly, X's system must be calibrated and the loudness control fitted to it. [14]

LOUDNESS VS. INTENSITY

The *phon* is a unit of loudness level and is tied to sound pressure level at 1,000 Hz as we have seen in Figs. 2-6, 2-7 and 2-8. It is useful, up to a point, but it tells us little about human reaction to loudness of sound. We need some sort of subjective unit of loudness. Many experiments conducted with hundreds of subjects and many types of sounds have yielded something of a consensus that for a 10 dB increase in sound pressure level the average person would say that loudness doubled. For a 10 dB decrease in sound level subjective loudness would be half. One researcher says this should be 6 dB, others 10 dB and work on the problem continues. However, a unit of subjective loudness has been adopted and it is called the *sone*. One sone is defined as the loudness experienced by a person listening to a tone of 40 phon loudness level. A sound of 2 sones is twice as loud, 0.5 sone half as loud. Table 2-1 gives the relationship between loudness level in phons to the subjective loudness in sones. Although most audio workers will have little occasion to become involved in phons or sones, it is well to realize that a true subjective unit of loudness (sone) is related to loudness level (phon) which is, in turn, related by definition to what we can measure with a sound level meter. There are highly developed empirical methods of calculating the loudness of sound as they would be perceived by humans from purely physical measurements of sound spectra as, for example, measured with a sound level meter and octave or one third octave filters.

LOUDNESS VS. BANDWIDTH

In our discussion of loudness we have talked tones up to this point, but single frequency tones do not give all the information we need to relate subjective loudness to meter readings. The noise of a jet aircraft taking off sounds much louder than a tone of the same sound pressure level. The bandwidth of the noise affects the loudness of the sound, at least within certain limits. Let's see how it does this and just what those limits are.

Figure 2-9A represents three sounds having the same sound pressure level of 60 dB. Their bandwidths are 100, 160, and 200 Hz, but heights (representing sound intensity per Hz) vary so that areas are equal. In other words, the three sounds have equal intensities. (Sound intensity has a specific meaning in acoustics

Table 2-1. The Relationship Between Loudness Level in Phons and Sones.

Loudness Level, phons	Subjective Loudness, sones	Typical examples
100	64	Heavy truck passing
80	16	Talking loudly
60	4	Talking softly
40	1	Quiet room
20	0.25	Very quiet studio

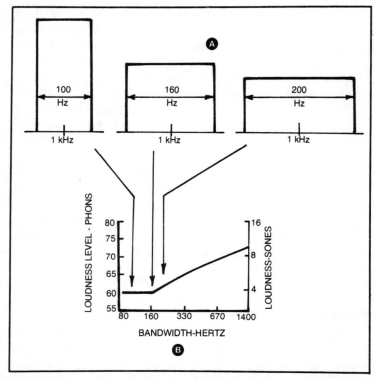

Fig. 2-9. (A) Three examples of noise of different bandwidths, but all having the same sound pressure level of 60 dB. (B) The loudness of the 100 and 160 Hz noises is the same, but the 200 Hz band sounds louder because it exceeds the 160 Hz critical bandwidth of the ear at 1,000 Hz. (Reference 17).

and is not to be equated to sound pressure. Sound intensity is proportional to the square of sound pressure for a plane progressive wave). The catch is that all three sounds of Fig. 2-9A do not have the same loudness. The graph in Fig. 2-9B shows how a bandwidth of noise having a constant 60 dB sound pressure level and centered on 1,000 Hz is related to loudness as experimentally determined.[17] The 100 Hz noise has a loudness level of 60 phons and a loudness of 4 sones. The 160 Hz bandwidth, it turns out, has the same loudness. But something mysterious happens as the bandwidth is increased beyond 160 Hz. The loudness of the noise of 200 Hz bandwidth is louder and from 160 Hz up, increasing bandwidth increases loudness. Why the sharp change at 160 Hz?

It turns out that 160 Hz is the width of the ear's *critical band* at 1,000 Hz. If a 1,000 Hz tone is presented to a listener along with random noise, only that noise in a band 160 Hz wide is effective in

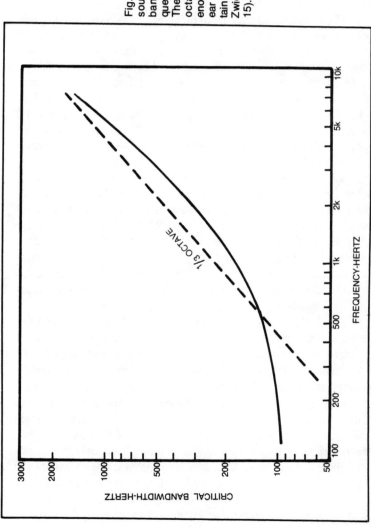

Fig. 2-10. The ear is basically a sound analyzer having critical bandwidths which vary with frequency according to the solid curve. The broken line shows that one third octave bandwidths are close enough to the critical bands of the ear to recommend their use in certain types of measurements (after Zwicker and Stevens, Reference 15).

masking the tone. In other words the ear acts like an analyzer composed of a set of filters adjacent to each other. The width of these critical bands varies with frequency as shown in Fig. 2-10[15,16] The bandwidth of one third octaves, plotted as a broken line, varies as a constant percentage of the center frequency (about 23%). The one third octave bandwidths approach the critical bandwidths of the ear closely enough to make them useful in certain loudness calculations and in other ways.

HEARING IMPULSES

The examples discussed to this point have considered steady state tones and noise. How does the ear respond to transients of short duration? This is important because music and speech are filled with transients, including assorted blasts, bangs, and pops. To focus attention on this aspect of speech and music play some tapes backward using the simple rethreading expedient of Fig. 2-11. The initial transients now appear at the end of syllables and musical notes and really stand out. These transients are everywhere and justify spending a few words on the ear's response to short lived sounds.

A 1,000 Hz tone sounds like 1,000 Hz in a 1 second tone burst, but an extremely short burst sounds like a click. Duration of such a burst also influences the perceived loudness. Short bursts do not sound as loud as longer ones. Figure 2-12 shows how much the level of shorter pulses have to be increased to have the same loudness as a long pulse or steady tone.[17] A pulse 3 milliseconds long must have a level about 15 dB higher to sound as loud as a 0.5 second (500 millisecond) pulse. Tones and random noise follow roughly the same relationship in loudness vs. pulse length.

The 100 ms region is significant in Fig. 2-12. Only when the tones or noise bursts are shorter than this amount must the sound pressure level be increased to produce a loudness equal to that of long pulses or steady tones or noise. This 100 ms appears to be the integrating time or the time constant of the human ear.

In reality Fig. 2-12 tells us that our ears are less sensitive to short transients. This has a direct bearing on understanding speech. It is well known that the consonants of speech determine the meaning of many words. For instance, the only difference between *bat, bad, back, bass, ban*, and *bath* are the consonants at the end. The words *led, red, shed, bed, fed*, and *wed* have the all important consonants at the beginning. No matter where they occur, these consonants are genuine transients having durations of the order of 5 to 15 ms. A glance at Fig. 2-12 tells us that transients this short must be of much higher level to be comparable to longer sounds. In the above words each consonant is not only much shorter than the rest of the word, they are also of lower level. This places a premium on having good listening conditions or transmission channels to distinguish between such words as the above sets. Such things as too high background noise or too much reverberation can cause serious reduction in the understandability of speech because of the consonant problem [10].

BINAURAL LOCALIZATION

The response of human ears to very short delays provides the basis for determining the direction from which sounds come. This ability to localize sound has been found to be keener for complex sounds than for tones. It is amazingly accurate in angular discrimination, 1° or 2°, when sounds arrive from directly in front

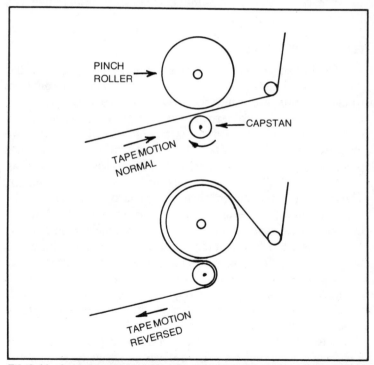

Fig. 2-11. A simple method of threading a tape recorder to reverse tape motion in an experiment to emphasize the important part transients play in music and speech. Only expendable tape should be used as the capstan bears on the coated side of the tape.

and at eye level. Considering sound sources located in the median plane, that vertical plane passing through the center of the head midway between the ears, there is a certain confusion between front and back. That is, sounds arriving at both ears simultaneously do not give definite cues as to location of the source. An absence of head shadow also contributes to the difficulty of locating sounds in the median plane [18] [146].

This lack of ability to localize sound sources in the median plane is a definite advantage in sound reinforcement work. There is confusion in having the visual image of a person speaking at the podium and the amplified sound of that person's speech coming from some other direction. This problem is especially aggravated by having a loudspeaker on each side of the platform. By locating a single loudspeaker cluster high over the podium, as heads of those in the audience turn toward the podium, the loudspeaker cluster is automatically placed in everyone's median plane. The confusion is thus reduced or eliminated as the source of sound is identified with the one at the podium.

PITCH VS. FREQUENCY

Pitch, a subjective term, is chiefly a function of frequency, but not linearly related to it. Because pitch is somewhat different from frequency, it requires a nice subjective unit and that unit is the *mel*. Frequency is a physical term measured in cycles per second, now called hertz. Although a weak 1,000 Hz signal is still 1,000 Hz if we increase its level, the pitch of a sound may depend on sound pressure level. A reference pitch of 1,000 mels has been defined as that pitch of a 1,000 Hz tone with a sound pressure level of 60 dB. The relationship between pitch and frequency, determined by experiments with juries of listeners, [19] is shown in Fig. 2-13. It is to be noted on the experimental curve that 1,000 mels coincides with 1,000 Hz which tells us that the sound pressure level for this curve is 60 dB. It is challenging to note that the shape of the curve of Fig. 2-13 is quite similar to a plot of position along the basilar membrane of the inner ear as a function of frequency. This suggests that pitch is related to action on this membrane, but much work remains to be done to be certain of this.

Intensity of sound has its effect on the perception of pitch. For low frequencies, the pitch goes down as intensity of sound is increased. At high frequencies the reverse takes place, the pitch increases with intensity. Fletcher reported an interesting illustration of this effect. Playing tones of 168 and 318 Hz at normal levels

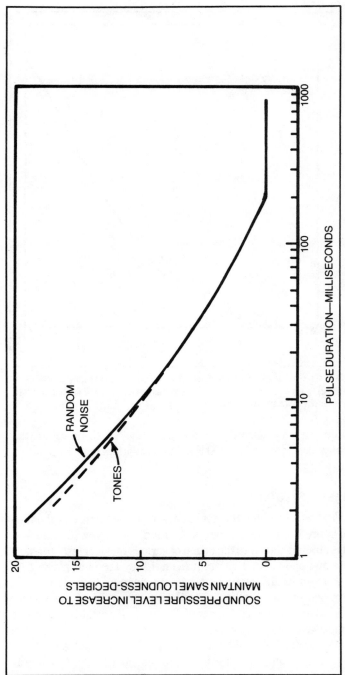

Fig. 2-12. Short pulses of tones or noises are less audible than longer pulses according to the above graphs. The discontinuity in the 100-200 ms region is related to the integrating time of the ear. (Reference 17)

45

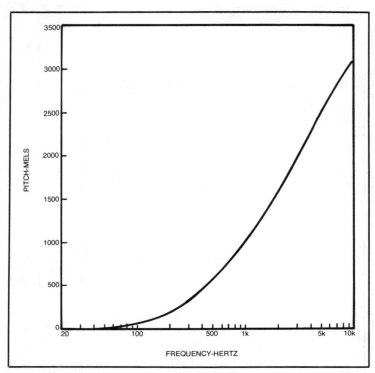

Fig. 2-13. Pitch (in mels, a subjective unit) is related to frequency (in hertz, a physical unit) according to the solid curve obtained by juries of listeners (after Stevens and Volkman, Reference 19).

resulted in a very discordant sound. At a high level, however, the ear hears them in the 150-300 Hz octave relationship as a pleasant sound.

TIMBRE VS. SPECTRUM

Timbre is not what the lumberjack yells as the tree is about to fall. Timbre has to do with our perception of complex sounds. The word is applied chiefly to the sound of various musical instruments. A flute and oboe sound different even if they are both playing A. The tone of each instrument has its own timbre. Timbre is determined by the number and relative strengths of the instrument's partials of the fundamental. Tonal quality would come close to being a synonym for timbre.

Obviously, timbre is very much another of those subjective terms. The analogous physical term is spectrum. A musical instrument produces a fundamental and a set of partials (or harmonics) which can be analyzed with a wave analyzer and plotted in the form of Fig. 1-7. Let us say that the fundamental is 200 Hz, second harmonic 400 Hz, third harmonic 600 Hz, etc. The subjective pitch which the ear associates with our measured 200 Hz., for example, varies with the level of the sound. The ear has its subjective interpretation of the harmonics also. Thus the ear's perception of the overall timbre of the instrument's note may be considerably different from the measured spectrum in a very complex way.

In listening to an orchestra while seated in a music hall the timbre one hears is different for different locations in the seating area.[20] The music is composed of a wide range of frequencies and the amplitude and phase of the various components are affected by reverberation. The only way to get one's analytical hands on studying such differences would be to study the sound spectra at different locations. But these are physical measurements and still the will-o-the-wisp of subjective timbre tends to slip away from us. The important point of this section is to realize that a difference exists between timbre and spectrum.

THE EAR IS NOT LINEAR

An ideal amplifier gives exactly twice the output if the input is doubled and no new frequency components are added, i.e., there is no distortion. There is no such amplifier, nor is the ear perfect in this regard. Here is an experiment suggested by Craig Stark[21] which can be easily performed with the home hi fi and two audio oscillators. Plug one oscillator into the left channel and the other into the right channel and adjust both channels for equal and comfortable level at some midband frequency. Set one oscillator to 24 kHz and the other to 23 kHz without changing level settings. With either oscillator alone, nothing is heard because the signal is outside the range of the ear (he notes here, however, that the dog may leave the room in disgust!). When both oscillators are feeding their respective channels, one at 24 kHz and the other at 23 kHz, one can hear a distinct 1,000 Hz tone if the tweeters are good enough and you are standing in the right place.

The 1,000 Hz tone is the difference between 24,000 and 23,000 Hz. The sum, or 47,000 Hz, which even the dog may not hear even if it were radiated, is another sideband. Such sum and difference sidebands are generated whenever two pure tones are mixed in a non-linear element. The non-linear element in the above experiment is the middle and inner ear. Sometimes these new frequencies generated are called beats. The effects of the ear's non-linearity is more pronounced at high sound levels. In addition to the intermodulation products discussed above, the non-linearity of the ear generates new harmonics not in the sound falling on the eardrum.

HAAS SENSE

Helmut Haas' name is immortalized in the *Haas Effect*, although many other investigators in the area preceded and followed him. Haas' work[22] certainly came at a time when audio people were open to the results of his work to explain some mystifying effects which had been observed. Haas pointed out that our hearing mechanism integrates the sound intensities over short intervals and compared it to a ballistic measuring instrument. In simpler terms in an auditorium situation the ear and brain has the remarkable ability of gathering together all reflections arriving within about 50 ms after the direct sound and combining (integrating) them and giving the impression that all this sound is from the direction of the original source even though reflections from other directions are involved. This integrated sound energy over this period also gives an impression of added loudness.

Haas set his subjects 3 meters from two loudspeakers arranged so that they subtended an angle of 45 degrees, the observer's line of symmetry splitting this angle. The conditions were approximately anechoic. The observers were called upon to adjust an attenuator until the sound from the "direct" loudspeaker was equal to that of the "delayed" loudspeaker. He then proceeded to study the effects of varying the delay.

A number of researchers had previously found that very short delays (less than 1 ms) were involved in our discerning the direction of a source by slightly different times of arrival at our two ears. Delays greater than this do not affect our directional sense.

As shown in Fig. 2-14, Haas found that in the 5-35 ms delay range the sound from the delayed loudspeaker had to be increased more than 10 dB over the direct before it sounded like an echo. This is the *precedence effect* or *Haas Effect*. In a room, reflected energy arriving at the ear within 50 ms is integrated with the direct sound and is perceived as part of the direct sound as opposed to reverberant sound. These early reflections increase the loudness of the sound and, as Haas has said, result in "a pleasant modification of the sound impression in the sense of broadening of the primary sound source while the echo source is not perceived acoustically". This is exactly the rationale behind the recent trend toward the live end - dead end approach to monitoring rooms (Chapter 16).

The transition zone between the integrating effect, for delays less than 50 ms, and the perception of delayed sound as an echo is gradual and therefore somewhat indefinite. Some peg the dividing line at a convenient 1/16th second (62 ms), some at 80 ms, and some at 100 ms beyond which there is no question of the discreteness of the echo. For the purposes of this book we shall consider the first 50 ms, as per Fig. 2-14, the region of definite integration. In Chapter 7 the subject of echoes will be treated further.

THE EAR AS A MEASURING INSTRUMENT

The emphasis on the distinction between physical measurements and subjective sensation would seem to rule out the possibility of the ear being used for physical measurements. True, we connot get digital readouts by looking in someone's eyes (or ears) but the ears are very keen in making comparisons. A person is able to detect differences in sound level of about 1 dB throughout

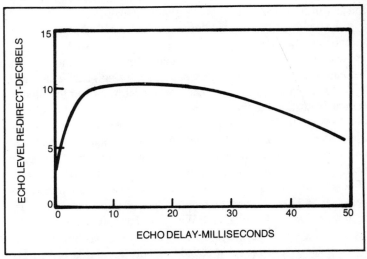

Fig. 2-14. The Haas Effect or precedence effect in the human ear. In the 5-30 ms region echo levels must be about 10 dB higher than the direct sound to be discernable as echoes. In this region of delay reflected components arriving from many directions are all combined by the ear making the sound louder and appear to come from the source. For delays greater than the 50-100 ms transition region reflections are perceived as discrete echoes (after Haas, Reference 22).

most of the audible band if the level is reasonable. Under ideal conditions a change of a third this amount is perceptible. As for detecting differences in frequency, at ordinary levels and for frequencies less than 1,000 Hz, the ear can tell the difference between tones separated by as little as 0.3%. This would be 0.3 Hz at 100 Hz and 3 Hz at 1,000 Hz.

The eminent Harvey Fletcher[23] has pointed out how the remarkable keenness of the human ears saved the day in many of his researches in synthesizing musical sounds. For example, in his study of piano sounds it was initially postulated that all that would be necessary would be to measure the frequency and magnitude of fundamental and harmonics and then combine them with the measured values of attack and decay. When this was done the listening jury unanimously voted that the synthetic sounds did not sound like piano sounds but more like organ tones. Further study revealed the long known fact that piano strings are stiff strings having properties of both solid rods and stretched strings. The effect of this is that piano partials *are non-harmonic*! By correcting the frequencies of what were assumed to be harmonics in integral multiples, the jury could not distinguish between the synthetic piano sounds and the real thing. The critical faculty of the ears of the jury in comparing sound qualities provided the key.

Knowledge of the ear's filter-like critical bands leads to the tantalizing idea of analyzing continuous noises such as traffic noises, underwater background noises, etc., by using the ear instead of heavy and expensive sound analyzing gear. This must have occurred to Harvey Fletcher, who first proposed the idea of critical bands, and to the many investigators in this field who have dealt with critical bands through the years. The general approach is illustrated in Fig. 2-15. A tape recording of the noise to be analyzed is played back and mixed with a tone from a variable frequency oscillator. The combination is amplified and listened to with a pair of headphones having a flat frequency response. The oscillator is set, say, at 1,000 Hz and its output adjusted until the tone is just hidden or masked by the noise. We know that only the noise in the critical band centered on 1,000 Hz is effective in masking the tone. If the noise is expressed in sound pressure level of a band 1 Hz wide, the voltage of the tone then corresponds to the 1 Hz sound pressure level of the noise at the masked point. By adjusting the voltage until the tone is just masked, we have the condition that should yield one point on our noise spectrum graph. For convenience, let us assume that this voltmeter is calibrated in dB referred to some arbitrary base such as 1 volt (dBv). Referring to Fig. 2-10 we note that the critical band centered on 1,000 Hz is 160 Hz wide. This can also be expressed in decibels by taking $10 \log_{10} 160 = 22$ dB. This 22 dB, representing the width of the critical band as it does, must be subtracted from the voltmeter reading in dB. This gives one point on the noise spectrum graph. Repeating the process for other frequencies a series of points is obtained which reveal the shape of the noise spectrum. If the recording and the entire measuring system (including the observer's ears) were calibrated, absolute levels for the noise spectrum could be obtained.

The interest here is that there is such a set of filters in our head which could be put to such a task, not that this method will ever replace a good sound level meter equipped with octave or one third octave filters. Surely human variabilities would far exceed sound level meter fluctuations from day to day and what the observer ate for breakfast has no effect on the sound level meter, although it might affect the dependability of the readings made with physiological equipment.

HEARING LOSS WITH AGE

The results of three separate studies on the average deterioration of hearing acuity with age are summarized[24] in Fig. 2-16 A and B. There seems to be a pronounced difference in this type of loss between men and women. It is not known whether the greater loss observed in large populations of men can be attributed, partially at least, to the noisier places in which men often work.

It is wise for an older person working in audio and dependent on keen ears for a livelihood to be aware of this irrevocable loss of

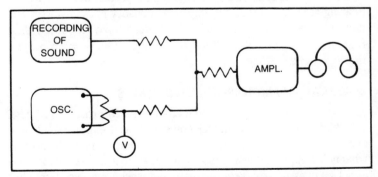

Fig. 2-15. Equipment arrangement for using critical bands of the ear for sound analysis.

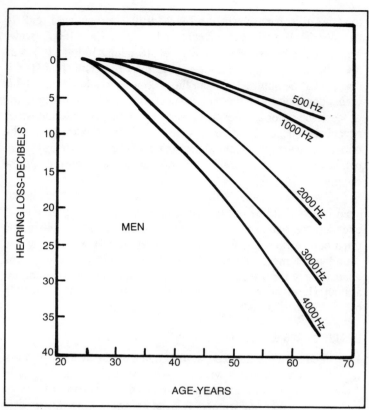

Fig. 2-16 A . Age related hearing loss of men (Reference 24).

high frequency sensitivity. By being aware, compensation can be made or, when it becomes a problem, a hearing aid can be fitted. Artists wear eyeglasses, why shouldn't a person in audio wear a hearing aid which results in improved sensory performance? The new electret microphones of such devices have smoothed their response considerably, although there is plenty of room for improvement.

OCCUPATIONAL AND RECREATIONAL DEAFNESS

The hearing of workers in industry is now protected by law. The higher the environmental noise, the less exposure allowed (Table 2-2). There is feverish activity today in trying to determine what noise exposure workers are subjected to in a given plant. This is not easy as noise levels fluctuate and workers move about, but wearable *dosimeters* are often used in integrate the exposure over the work day. Industries are hard pressed to keep up with changes in regulations, let alone the installation of noise shields around offending equipment and keeping ear plugs in or ear muffs on the workers. Nerve deafness resulting from occupational noise is recognized as a distinct health hazard.

It is especially bad when one working all day in a high noise environment then engages in motorcycle or automobile racing, listens to a 400 watt stereo at high level, or spends hours in a discotheque. The professional audio engineer operating with high monitoring levels is risking irreparable injury to the basic tools of the trade . . . ears. As high frequency loss creeps in the volume control is turned up to compensate and the rate of deterioration is accelerated.

Is this concern well founded or a case of crying wolf? Some rather startling data have resulted from studies on students. Raichel[25] summarizes several studies. In one case it was found that 3.8% of sixth graders failed a high frequency hearing test, but 11% of ninth graders and 10.6% of high school seniors failed it. A survey of incoming college freshman yielded a 33% rate of failure. The next year 60.7% of the new incoming class failed. Specialists consider nerve deafness the leading disability in the United States.

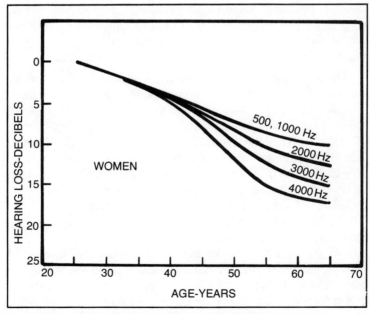

Fig. 2-16B. Age related hearing loss of women (Reference 24).

53

Table 2-2. OSHA Permissible Noise Exposure Times.*

Sound Pressure Level, dB, A-Weighting, Slow Response	Maximum Daily Exposure Hours
85	16
90	8
92	6
95	4
97	3
100	2
102	1.5
105	1
110	0.5
115	0.25 or less

*Reference 26

As there is no replacement known for damaged sensory cells of the inner ear, it would seem wise to protect the ones we have. [27]

ZUSAMMENFASSUNG

The word *summary,* which implies a brief recapitulation, is not as suitable in this case as the German word for summary, *zusammenfassung,* which literally means holding together. This chapter on ear characteristics of interest to audio workers has gotten somewhat out of hand and definitely needs to be held together. Here are some key points worthy of a second look:

■ The ear is sensitive enough to hear the tattoo of air particles on our eardrums in the quietness of an anechoic chamber.

■ The auditory canal, acting as a quarter wave pipe closed at one end with the eardrum, contributes an acoustical amplification of about 10 dB and the head diffraction effect another 10 dB near 3 kHz. These are vital speech frequencies.

■ The leverage of the ossicle bones of the middle ear and the ratio of areas of eardrum and oval window successfully match the impedance of tenuous air to the liquid of the inner ear.

■ The Eustachian tube and round window provide pressure release and equalization with atmospheric pressure.

■ Waves set up in the inner ear by vibration of the oval window excite the sensory hair cells which are connected to the brain. There is a "place effect", the peak of hair cell agitation for higher frequencies being nearer the oval window, low frequencies at the far end. These "tuning curves" are very broad. The observed sharpness of the ear's ability to analyze sound must be the result of neural function.

■ The area of audibility is proscribed by two threshold curves, the threshold of audibility at the lower extreme and the threshold of feeling or pain at the loud extreme. Our entire auditory experience is played out within these two extremes.

■ The equal loudness contours of Fig. 2-7 are, in effect, inverted frequency response curves of the human ear. The shapes of these curves vary with loudness level, becoming flatter at high levels.

■ Loudness of wideband noise (assuming constant band level) does not increase with increase in bandwidth until the critical bandwidth of the ear is exceeded.

■ The critical bands of the ear make the ear an analyzer of sound. These bands are approximated by one third octaves.

■ The loudness of tone bursts decreases as the length of the burst is decreased. Bursts greater than 100-200 ms have full loudness, indicating a time constant of the ear near 100 ms.

■ Our two ears are capable of accurately locating the direction of a source in the horizontal plane. In a vertical median plane, however, locating ability is quite poor. This means that a sound reinforcing loudspeaker above the one talking gives the illusion of all sound coming from the person talking.

■ Pitch is a subjective term. Frequency is the associated physical term and the two have only a general relationship.

■ Subjective timbre or quality of sound and the physical spectrum of the sound are related, but not equal.

■ The non-linearity of the ear generates intermodulation products and spurious harmonics.

■ The Haas or precedence effect describes that effect of the ear which integrates all sound arriving within the first 50 ms, making it sound louder.

■ Although the ear is not effective as a measuring instrument yielding absolute values, it is very keen in comparing frequencies, levels, or harmonic content.

■ A progressive high frequency hearing loss is the prospect for all as part of the aging process.

■ Occupational and recreational noises are taking their toll in permanent hearing loss. As opposed to hearing loss due to aging, we can take definite precautionary steps to minimize this type of environmentally caused deafness.

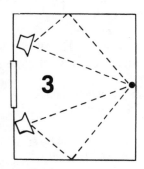

3 Levels and the Decibel

The decibel is as common a unit in audio circles as a minute or mile is in common usage. The decibel is a unit of many kinds of levels used in acoustics. The way the dB is tossed around one gets the impression that it is as well understood as a mile or a minute. With a little thought this can be the case.

A level in decibels makes it easy to handle measurements pertaining to the human senses. The popularity of the decibel is really a tribute to the extremely wide ranges involved in the sensitivity of our senses. In Chapter 2 we were given cause to stand in awe at the tremendous range of sensitivity of our sense of hearing, but other senses display remarkably wide ranges as well. We saw that the threshold of hearing matches the ultimate lower limit of perceptible sound in air, the noise of air molecules striking the eardrum. The visual threshold is no less amazing. The sensitivity of normal human eyes also matches the ultimate limit by responding to one or a very few photons of light. From these threshold responses to the most feeble stimuli, the ear and the eye are also capable of handling high intensities of sound and light. A level in decibels is a convenient way of handling the billion-fold (10^9) range of sound intensity to which the ear is sensitive, without causing us to stumble over long strings of zeros.

RATIOS VS. DIFFERENCES

Let us imagine a sound source set up in front of us as we carry out this imaginary experiment in a room completely protected from interfering sounds. (The term "sound proof room" was not used because we intend to have plenty of sound in it.) The sound source is adjusted for a very weak sound having an intensity of 1 unit and we note carefully its loudness. When the intensity is increased until it sounds twice as loud the intensity dial is noted to read about 10 units. This completes observation A. For observation B the source intensity is increased to 10,000 units. To double the loudness we find that the intensity must be increased from 10,000 to 100,000 units. The results of this experiment may now be summarized as follows:

	Differences between two intensities	Ratio of two intensities
Observation A	$10 - 1 = 9$	10 to 1
Observation B	$100,000 - 10,000 = 90,000$	10 to 1

Observations A and B accomplished the same thing, doubling the perceived loudness. In observation A this was accomplished by an increase in intensity of only 9 units, while in observation B it took 90,000 units. Ratios of intensities seem to describe loudness changes better than differences of intensity. Weber (1834), Fechner (1860), Helhholtz (1873), and other early researchers pointed out the importance of ratios, which we know apply equally well to vision, hearing, vibration, or even electric shock.

Many years ago a friend demonstrated his experiment on the hearing of cats which, in many ways, is similar to that of humans. A tone of 250 Hz, radiated from a nearby loudspeaker, was picked up by the ears of the anesthetized cat, a portion of whose brain was temporarily exposed. A delicate probe picked up the 250 Hz signal at a highly localized spot on the acoustic cortex, displaying it on a cathode ray oscilliscope. When the tone was shifted to 500 Hz, the signal was picked up at another spot on the cortex. Tones of 1,000 and 2,000 Hz were detected at other specific spots. The fascinating point here is that changing the tone an octave resulted in the signal appearing on the acoustic cortex at discrete equally spaced points. Frequencies in the ratio of 2 to 1 (an octave) seem to have a linear positional relationship in the cat's brain. Ratios of stimuli come closer to matching up with human perception than do differences of stimuli. This matching is not perfect, but close enough to make a strong case for the use of levels in decibels.

HANDLING NUMBERS

Table 3-1 illustrates three different ways numbers may be expressed. We are familiar with the decimal and arithmetic forms

Table 3-1. Ways of Expressing Numbers.

Decimal Form	Arithmetic Form	Exponential Form
100,000	$10 \times 10 \times 10 \times 10 \times 10$	10^5
10,000	$10 \times 10 \times 10 \times 10$	10^4
1,000	$10 \times 10 \times 10$	10^3
100	10×10	10^2
10	10×1	10^1
1	$10/10$	10^{10}
0.1	$1/10$	10^{-1}
0.01	$1/(10 \times 10)$	10^{-2}
0.001	$1/(10 \times 10 \times 10)$	10^{-3}
0.0001	$1/(10 \times 10 \times 10 \times 10)$	10^{-4}
100,000	$(100)(1,000)$	$10^2 + 10^3 = 10^{2+3} = 10^5$
100	$10,000/100$	$10^4/10^2 = 10^{4-2} = 10^2$
10	$100,000/10,000$	$10^5/10^4 = 10^{5-4} = 10^1 = 10$
10	$\sqrt{100} = \sqrt[2]{100}$	$100^{1/2} = 100^{0.5}$
4.6416	$\sqrt[3]{100}$	$100^{1/3} = 100^{0.333}$
31.6228	$\sqrt[4]{100^3}$	$100^{3/4} = 100^{0.75}$

in everyday activity. The exponential form, while not as commonly used, has the charm of simplifying things once we conquer the fear of the unknown or little understood. In writing one hundred thousand, we may have a choice between 100,000 watts and 10^5 watts, but how about a millionth of a millionth of a watt? All those zeros behind the decimal point make it impractical even to reproduce here, but 10^{-12} is easy. And the single word that means 10^{-12} is pico; so the power is one picowatt. Engineering type calculators care for the exponential form in what is called "scientific notation" by which very large or very small numbers may be entered.

Representing 100 as 10^2 simply means that $10 \times 10 = 100$ and that 10^3 means $10 \times 10 \times 10 = 1,000$. But how about 267? Voila! That is where logarithms come in. We agree that $100 = 10^2$. By definition we can say that the logarithm of 100 to the base $10 = 2$, commonly written $\log_{10} 100 = 2$ or simply $\log 100 = 2$, because common logarithms are to the base 10. Now that number 267 above needn't scare us, it is simply expressed as 10 to some other power between 2 and 3. The old fashioned way was to go to a book of log tables but with a simple handheld calculator we punch in 267 and push the "log" button and 2.4265 appears. Thus $267 = 10^{2.4265}$ and $\log 267 = 2.4265$. Logs are so handy because, as Table 3-1 demonstrates, they reduce multiplication to addition, division to subtraction. This is exactly how the now extinct slide rule or "slipstick" worked. Logs should be the friend of every audio worker for they are the solid foundation of our levels in decibels. In fact, a level is a logarithm of a ratio. A level in decibels is ten times the logarithm to the base 10 of the ratio of two power like quantities.

BACK TO RATIOS

Ratios of powers or ratios of intensities or ratios of sound pressure, voltage, current, or anything else are dimensionless. For instance, the ratio of 1 watt to 100 watts is 1 watt/100 watts and the watt unit upstairs and the watt unit downstairs cancel leaving $1/100 = 0.01$, a pure number without any watt dimension. This is important for logarithms can be taken only of nondimensional numbers.

DECIBELS

A power level may be expressed in bels (from Alexander Graham Bell):

$$\log_{10} \frac{W_1}{W_2} \tag{3-1}$$

Because the decibel, from its very name, is 1/10 of a bel, the level in decibels of a power ratio becomes:

$$L_1 = 10 \log_{10} \frac{W_1}{W_2} \tag{3-2}$$

Equation 3-2 applies equally to acoustic power, electric power, or any other kind of power. A question often arises when levels other than power need to be expressed in decibels. For example, acoustic intensity is acoustic power per unit area in a specified direction, hence Equation 3-2 can be used for both. Acoustic power is proportional to the square of the acoustic pressure and:

$$\begin{aligned} L_p &= 10 \log \frac{p_1^2}{p_2^2} \\ &= 20 \log \frac{p_1}{p_2} \quad \text{in decibels} \end{aligned} \tag{3-3}$$

The tabulation of Table 3-2 will help in deciding whether the Equation 3-2 or Equation 3-3 form applies.

Sound pressure is usually the most accessible parameter to measure in acoustics, even as voltage is for electronic circuits. For this reason the Equation 3-3 form is more often encountered in day to day technical work.

Table 3-2. Use of 10 Log and 20 Log.

	Eq (3-2) $10 \log_{10} \dfrac{a_1}{a_2}$	Eq (3-3) $20 \log_{10} \dfrac{b_1}{b_2}$
ACOUSTIC		
Power	X	
Intensity	X	
Air particle velocity		X
Pressure		X
ELECTRIC		
Power	X	
Current		X
Voltage		X

REFERENCE LEVELS

With a sound level meter a certain sound pressure level is read. If the corresponding sound pressure is expressed in normal pressure units we are confronted with a great range of very large and very small numbers. Ratios are more closely related to human senses than linear numbers and "the level in decibels approach" compresses the large and small ratios into a more convenient and comprehensible range. Basically our sound level meter reading is a certain sound pressure level, $20 \log (p_1/p_2)$ as in Equation 3-3. We need some standard reference sound pressure p_2. It is important that the reference p_2 selected be the same as that used by others, so ready comparisons can be made worldwide. Several such reference pressures have been used over the years but for sound in air the standard reference pressure has been 20 μPa (micropascal). This may seem quite different from the reference pressure of 0.0002 microbar or 0.0002 dyne/cm^2, but it is the same standard merely written in different units. This is a very minute sound pressure which corresponds closely to the threshold of human hearing.

When a statement is encountered such as, "The sound pressure *level* is 82 dB", we are normally content to use the 82 dB sound pressure level directly in comparison with other levels. But if the sound pressure were needed it can be readily computed by working backward from Equation 3-3 as follows:

$$82 = 20 \log \frac{p_1}{20 \, \mu \text{Pa}}$$

$$\log \frac{p_1}{20 \, \mu \text{Pa}} = \frac{82}{20}$$

$$\frac{p_1}{20\mu P_a} = 10^{\frac{82}{20}}$$
$$p_1 = (20\,\mu\text{Pa})\,(10^{4.1})$$

The y^x button on the calculator helps us to evaluate $10^{4.1}$. Enter 10; press y^x button; enter 4.1.* The answer 12,589 appears.

$$p_1 = (20\,\mu\text{Pa})\,(12,589.)$$
$$p_1 = 251,785.\,\mu\text{Pa}$$

No one will deny that "82 dB SPL" is easier to handle than sound pressure of 251,785. μPa.

There is another lesson here. The 82 has what is called two significant figures. The 251,785. has six and implies a precision that just is not there. Just because the calculator says so doesn't make it so!

Remember that sound pressure *level* in air means that the reference pressure downstairs (p_2) in the pressure ratio is 20 μPa. There are other reference quantities, some of the more commonly used are listed in Table 3-3. In dealing with very large and very small numbers a familiarity with the prefixes of Table 3-4 is advised. These prefixes are nothing more than Greek names for the powers (exponents) of 10.

ACOUSTIC POWER

It doesn't take many watts of acoustic power to produce some very loud sounds, as anyone who lives downstairs from a dedicated audiophile will testify. We are conditioned by megawatt electrical generating plants, 350 horsepower (261 kilowatt) automobile

Table 3-3. Reference Quantities in Common Use.

LEVEL IN DECIBELS	REFERENCE QUANTITY
ACOUSTIC	
Sound pressure level in air (SPL, dB)	20 micropascal
Power level (L_p, dB)	1 picowatt (10^{-12} watt)
ELECTRIC	
Power level re 1 mW	10^{-3} watt (1 milliwatt)
Voltage level re 1 V	1 volt
Volume level, VU	10^{-3} watt

*The order of entry is different with RPN calculators.

Table 3-4. The Greeks Had A Word For It.

Prefix	Symbol	Multiple
terra	T	10^{12}
giga	G	10^{9}
mega	M	10^{6}
kilo	k	10^{3}
milli	m	10^{-3}
micro	μ	10^{-6}
nano	n	10^{-9}
pico	p	10^{-12}

engines, and 1500 watt flatirons which eclipse the puny watt or so the hi fi loudspeakers may radiate as acoustic power. Even though a hundred-watt amplifier may be driving the loudspeakers, loudspeaker efficiency (output for a given input) is very low and headroom must be reserved for the occasional peaks of music. Increasing power to achieve greater results is often frustrating. Doubling power from 1 to 2 watts is a 3 dB increase in power level ($10 \log 2 = 3.01$); the same increase in level is represented by an increase in power from 100 to 200 watts or 1000 to 2000 watts.

Table 3-5 lists sound pressure and sound pressure levels of some common sounds. In the sound pressure column it is a long stretch from 100,000 Pa (10 kPa) which is atmospheric pressure to 0.00002 Pa (20 μPa), but this range is reduced to quite convenient form in the level column.

Another way to generate 194-dB sound pressure level, besides launching a saturn rocket, is to detonate 50 pounds of TNT 10 feet away. We saw that common sound waves are but tiny ripples on the steady-state atmospheric pressure. A sound of 194 dB sound pressure level approaches atmospheric and hence is a ripple of the same order of magnitude as atmospheric pressure. The 194 dB sound pressure is an rms (root mean square) value. A peak sound pressure 1.4 times as great would modulate the atmospheric pressure completely.

USING DECIBELS

The decibel as a unit of level can be handled like other units such as miles, feet, seconds, gallons, or pounds. A level is a logarithm of a ratio of two power-like quantities. When levels are computed from other than power ratios, certain conventions are involved. The convention for Equation 3-3 is that sound power is proportional to (sound pressure)^2. The voltage level gain of an

Table 3-5. Some Common Sound Pressure Levels and Sound Pressures.

Sound Source	Sound Pressure (Pa)	Sound Pressure Level* (decibels, A-weighted)
Saturn rocket	100,000. (one atmosphere)	194
Ram jet	2,000.	160
Propeller aircraft	200.	140
Threshold of pain		135
Riveter	20.	120
Heavy truck	2.	100
Noisy office, } Heavy traffic	0.2	80
Conversational speech	0.02	60
Private office		50
Quiet residence	0.002	40
Recording studio		30
Leaves rustling	0.0002	20
Hearing threshold, good ears at frequency of maximum sensitivity		10
Hearing threshold, excellent ears at frequency of maximum response	0.00002	0

* Reference pressure (take your pick, these are identical): 20 micropascal (μPa)	0.00002 pascal 2×10^{-5} newton/meter^2 0.0002 dyne/cm^2 or microbar

amplifier in decibels is 20 log (output voltage/input voltage); this holds regardless of the input and output impedances. However, for power-level gain the impedances must be considered if they are different. If it is a line amplifier with 600 ohm input and output impedances, well and good. Otherwise a correction is required. The important lesson here is to indicate clearly what kind of level is intended, or else label the gain in level as "relative gain, dB".

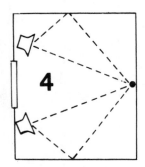

4   Sound Waves and the Great Outdoors

To understand complex acoustical phenomena, it helps to master the simple ones. An understanding of the various things that can happen to sound waves is basic to an understanding of complex sound fields indoors. The simplest effects are associated with a free field in which sound may travel unimpeded and unaltered in all directions. Even being outdoors may not provide a free field because the earth's surface is a reflecting surface, although certain experiments may be conducted in which such reflections are not a problem. For example, small manufacturers of loudspeakers, not able to afford an anechoic room, have dug a pit, mounted the loudspeaker under test face up and flush with the ground, and explored its directional and other characteristics by means of a microphone mounted high above on a flimsy jury-rig structure. Even here the earth plane acts as an infinite baffle. If the lougspeaker axis were parallel to the ground surface and close to it, the component of sound traveling directly to the microphone would interact with the component reflected from the earth's surface, greatly complicating the results. In this chapter we will, in most cases, stay far enough away from the earth's surface so that its effects can be disregarded.

SPREADING OF SOUND WAVES

In a free field an ideal point source of acoustical energy sends out sound of uniform intensity (power per unit area) in all directions. Measuring the sound intensity at successively increasing distances from the source, it is observed that the intensity decreases with distance. Now, this isn't really too surprising. We note that doubling the distance reduces the intensity to 1/4th the initial value, tripling the distance yields 1/9th, and increasing the distance four times yields 1/16th of the initial intensity. Intensity of sound, we find, is inversely proportional to the square of the distance. An examination of Fig. 4-1 reveals that the same energy travels through the area at distance d as successive segments of spherical surfaces at distances of 2d, 3d, and 4d. This illustrates the *inverse square law*.

But sound intensity is a very difficult thing to measure. Our common instruments measure sound pressure or, more commonly, sound pressure level. In Chapter 3 we saw that sound intensity is proportional to the square of the sound pressure. When considering sound pressure, then, the inverse square law becomes simply the *inverse distance law*. In other words, sound pressure varies inversely as the first power of distance. In Fig. 4-2 sound pressure level in decibels is plotted against distance. This illustrates the basis for the common and very useful expression, *6 dB decrease of sound pressure level per doubling of the distance*, which applies only for a free field. As we shall see later, a free field exists even in enclosed spaces under very limited circumstances.

When the sound pressure level L_1 at distance d_1 from a point source is known, the sound pressure level L_2 at another distance d_2 can be calculated from

$$L_2 = L_1 - 20 \log \frac{d_2}{d_1} \quad , \text{decibels} \qquad (4\text{-}1)$$

In other words, the difference in sound pressure level between two points which are d_1 and d_2 distance from the source is

$$L_2 - L_1 = 20 \log \frac{d_2}{d_1} \quad , \text{decibels} \qquad (4\text{-}2)$$

For example, if a sound pressure level of 80 dB is measured at 10 ft., what is the level at 15 ft.? Solution:
$20 \log 10/15 = -3.5$ dB; the level is $80 - 3.5 = 76.5$ dB.
What is the sound pressure level at 7 ft.? Solution:
$20 \log 10/7 = +3.1$ dB and level is $80 + 3.1 = 83.1$ dB. All this is for free field in which sound diverges spherically, but this procedure may be helpful for rough estimates even under other conditions.

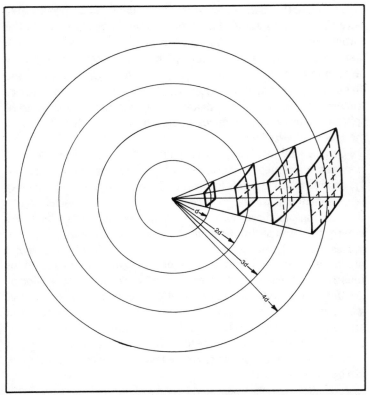

Fig. 4-1. In the solid angle shown, the same sound energy is distributed over spherical surfaces of increasing area as d is increased. The intensity of sound is inversely proportional to the square of the distance from the point source.

ABSORPTION OF SOUND IN AIR

The "loss" due to spherical divergence (it isn't a loss in the true sense) is so large it cannot very well be hidden. We simply sense that sound gets weaker as distance is increased. Not so with the absorption loss in air (a true loss) as it is of modest magnitude at modest distances for audio frequencies. Therefore, it may come as a surprise to some that there is any absorption of sound in the air. This true absorption loss of sound pressure level increases in proportion to distance traveled; it is greatest at higher audio frequencies. The absorption loss depends primarily on relative humidity of air, although air temperature has its effect as well. Figure 4-3 displays the dependence of absorption loss on frequency and relative humidity[28]. The loss at 10 kHz for 20% relative humidity is about 9 dB per hundred feet of travel. In large auditoriums or outdoor theaters this high frequency loss resulting from absorption in the air must be considered.

REFLECTION OF SOUND

In radio-frequency transmission lines any discontinuity sends reflected energy back toward the source. In an air conditioning system, equipment noise transmitted down a duct is partially reflected back toward its point of origin as it reaches the abrupt transition from duct to open air at the grille. We are not surprised when sound energy in air is reflected when it strikes hard objects, whether a yodel in the Alps or the echo from the whistle blasts that the captain of a fog-bound coastal steamer may release to gauge his distance from the cliffs.

If plane waves of sound strike a flat wall at normal (perpendicular) incidence, the reflection returns to the source. In Fig. 4-4 is illustrated the well-known fact that the angle of incidence a_i is equal to the angle of reflection a_r. Even for normal incidence this rule holds as a_i and a_r are both equal to zero.

In Fig. 4-5 sound falls on solid cylindrical surfaces. Plane waves are scattered widely when they strike the convex surface of

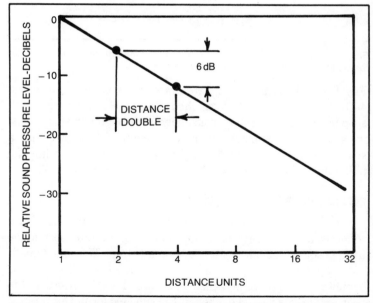

Fig. 4-2. When considering sound pressure, the inverse square law of intensity becomes simply the inverse distance law of sound pressure meaning that the sound pressure level is reduced 6 dB for each doubling of the distance.

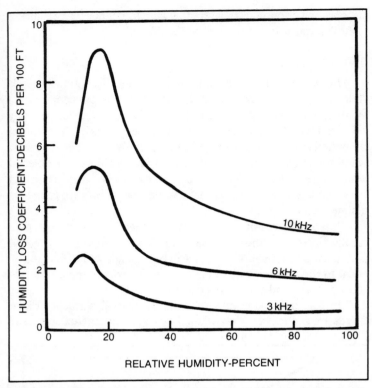

Fig. 4-3. Attenuation of a plane wave propagated in the atmosphere attributed to losses in the air (after Knudsen, Haris, Ref. 28).

Fig. 4-5A. The same cylindrical surface in concave form, as in Fig. 4-5B, however, tends to focus the reflected sound. Objects of all shapes reflect appreciable sound only when the dimensions of the irregularity are comparable to the wavelength of the sound. Thus a sea wall or vertical stone cliff might reflect ocean waves while the piling supporting a pier would have no detectable effect.

REFRACTION (BENDING) OF SOUND

In Fig. 4-4 the assumption was made that *all* the sound was reflected from the surface. Although this may be quite close to the truth in some cases, it is never completely true. The sound that is not reflected from the surface is transmitted into the material behind the surface. It may then be absorbed within, translated to heat, or emerge from the other side in subdued form. Our interest at the moment concentrates on the bending of the path of the sound ray.

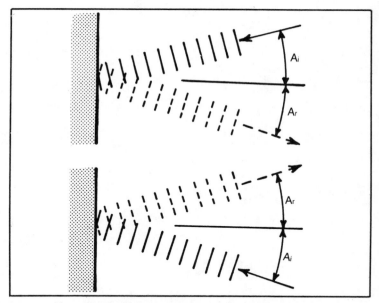

Fig. 4-4. The angle of incidence a_i is equal to the angle of reflection a_r.

The speed of sound depends upon the density of the medium. The speed of sound in tenuous air, as shown in Table 4-1, is 1130 ft. per second, but to be precise it is necessary to specify the air temperature. The speed of sound in several other common materials is listed in Table 4-1. Figure 4-6A depicts plane wave sound traveling from right to left, from a less dense medium to a denser medium, the latter suggested by the shaded area. The wave front is tilted by the denser medium. This is directly due to the higher speed of sound in the denser medium. Let us imagine that the lines of Fig. 4-6 represent ranks of soldiers marching at a uniform rate. The white area represents, in this wild example, a ploughed field which slows their gait and the shaded area represents a paved parade ground on which they can walk faster. Let us look carefully at what happens to the row of soldiers labelled pq. As soon as the soldiers on the p end of the row emerge from the ploughed ground, their speed increases, even though the cadence is the same. The soldiers on the q end of the row reach the hard surface later. This results in the entire company marching off in a new direction which the drill sergeant might call "left oblique". It all came about by the difference in walking speed between the white and shaded areas and the fact that the soldiers in each row arrived at the boundary at different times.

69

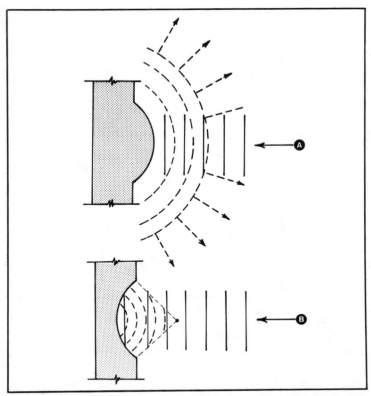

Fig. 4-5. (A) Plane waves of sound impinging on convex irregularities tend to be dispersed through a wide angle. (B) Concave surfaces tend to focus sound. Such irregularities, however, must have dimensions comparable to the wavelength of the sound to be effective.

In Fig. 4-6B we have the identical situation, but reversed, the column of solders marching from the parade ground onto the ploughed field. The same alteration in each row of soldiers occurs, but in the opposite direction.

The rows of soldiers are analogous to the plane wave fronts of sound. The ploughed ground (white area) represents a medium of low sound speed, the paved parade ground (shaded area) a medium of higher sound speed. This analogy helps one to figure out which way the sound ray will be bent, knowing which of the two media has the higher speed of sound and which the lower. This effect, called *refraction* of sound, is always present when media of different densities, and hence sound speeds, are encountered.

Pausing for a breather, what happens when the row of soldiers arrive at the boundary in Fig. 4-6A simultaneously? In this case all the soldiers speed up together and the spacing of the rows on the parade ground is increased. When soldier walking speed is increased, row spacing is increased. Equation 1-1 in Chapter 1 told us that the wavelength of sound is directly proportional to the speed of sound and we have also seen in Table 4-1 that the denser the medium, the greater the speed of sound. Our analogy rings true here, too, if the spacing between rows of soldiers is taken as wavelength.

Sound travels more rapidly in warm air than in cool air. If air temperature decreases with altitude, the speed of sound also decreases with altitude. This results in an upward bending of sound rays as illustrated in Fig. 4-7A. If the source of sound is high above the ground, such as an airplane, a shadow zone may be created. This represents the normal situation in which the heat of the sun warming the earth also warms the air near the earth. Figure 4-7B represents conditions which commonly prevail in the evening or during the night as the earth cools off after the sun sets. This accounts for being able to hear nighttime sounds at much greater distances than in the daytime because the sound rays are refracted earthward.

In our great outdoors we now go deep into the sea as we consider the example of sound refraction of Fig. 4-7C. Due to the changes of water temperature and pressure with depth, a sound velocity gradient of "V" shape occurs which creates a sound channel at a depth of approximately 4,000 feet. The sound of a depth charge detonated in this channel does not experience spherical divergence because refraction keeps the sound energy in the channel, spreading only horizontally. As sound tends to go upward, it is refracted back into the channel; as it tends to go downward it is refracted upward toward the center of the channel. Such channel borne sounds can be detected thousands of miles from the source and have been used to triangulate on downed aviators in life rafts.

Table 4-1. Speed of Sound.

Medium	Speed of Sound	
	Ft/Sec	Meters/Sec
Air	1,130	344
Sea water	4,900	1,500
Wood, fir	12,500	3,800
Steel bar	16,600	5,050
Gypsum board	22,300	6,800

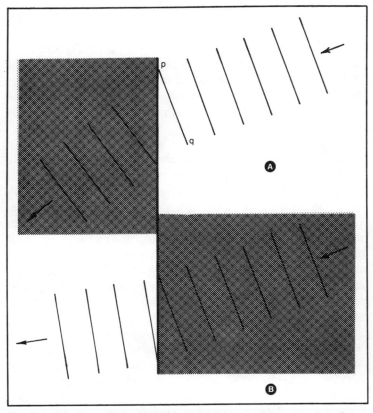

Fig. 4-6. The speed of sound is greater in a dense medium than a less dense one. A sound wave traveling from a medium of one density to another will be refracted from its original direction. (A) A plane wave traveling from air, for example, into a dense medium will have its direction refracted downward in the sketch. (B) The direction of a plane wave emerging from a dense medium into a less dense medium such as air will be bent upward.

DIFFRACTION OF SOUND

Refraction and diffraction sound very much alike and are easily confused. To add to the confusion, both effects result in a change in direction of the sound. However, the two mechanisms are entirely different. Refraction changes the direction of sound by virtue of different sound speeds in different media or velocity gradients in the same medium (Fig. 4-6 and 4-7). Diffraction is a change of direction resulting from passage around an obstacle. To illustrate this, let us consider sound traveling freely in our great outdoors striking the edge of an obstacle such as a high brick wall. We sense that practically no sound will penetrate the wall. Will there be any

72

sound behind the wall? The answer is "yes", because of this diffraction effect.

A plane wave train strikes the brick wall of Fig. 4-8A. The edge of the wall acts as a new source of sound, radiating into the shadow zone behind the wall. The sound is of a lower level, but it is there. If we take the distance between the lines representing the incident plane wave of Fig. 4-8A as the wavelength of the sound we note that the wall is large compared to the wavelength. For sound of very long wavelength, the same wall would be no barrier at all - the sound would sweep past as though the wall didn't exist. But for walls large in terms of the wavelength, this diffraction effect is responsible for sound penetrating the shadow zone behind the wall.

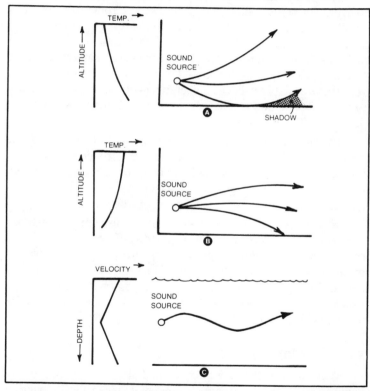

Fig. 4-7. (A) During the day while the sun is heating the earth the warm air below and the cool air above refracts sound rays upward, often creating a distant shadow zone. (B) At night the temperature gradient may be reversed causing sound to be refracted downward. This accounts for hearing sounds from great distances at night. (C) A V-shaped sound velocity gradient in the ocean creates a sound channel at about 4,000 foot depth, resulting in long range sound transmission.

In Fig. 4-8B the plane waves striking the steel plate with a hole in it are changed to spherical waves on the other side. This effect is dominant if the hole or slit is small compared to the wavelength of the sound falling on the plate. For very long wavelengths, the diffraction effect if small. Does diffraction have any real life application in audio? Consider the following:

■ Small cracks around observation windows, or back-to-back microphone or electrical service boxes in partitions can destroy the hoped for sound isolation between studios or between studio and control room. The sound emerging on the other side of the hole or slit is spread in all directions by diffraction.

■ The diffraction grating principle applies to sound as well as to light. The reflection of short, sharp sounds from a flight of steps or other periodic structure such as railing, or wooden fence can have a musical pitch. All those heads in neat rows in an auditorium may result in abnormal sound transmission characteristics at grazing angles due to diffraction. Periodic ornamentation such as architectural slats can have similar effects degrading sound quality to the audience or for microphone pickup. Beware of extensive periodic reflecting surfaces.

■ The human head is a diffracting obstacle to sound and speech directivity is thus dependent on frequency. At low frequencies a person speaking lays down an essentially nondirectional sound field. For the shorter wavelengths of higher frequencies the sound from the one speaking is quite directional.

■ The application of acoustical material in a studio in patches contributes to diffraction, scattering, and a diffusion of sound in the room. Repetitive, periodic patches, however, should give way to more random arrangement to avoid exceptional effects at one certain frequency determined by the spacing of the patches. For example, applying acoustical tile 12" x 12" in a checkerboard pattern is ill advised.

SUPERPOSITION OF SOUND

A sound contractor was concerned about the aiming of his horns in a certain auditorium. The simplest mechanical mounting would cause the beam of one horn to cut across the beam of the other horn. What happens in that bit of space where the two beams intersect? Would the beams tend to spread out? Would sound energy be lost from the beams as one beam interacts with the other? Relax. Nothing happens.

In the physics lab a large, but shallow, ripple tank of water is on the lecture table. The instructor positions three students around the tank, directing them to drop stones in the tank simultaneously. Each stone causes circular ripples to flow out from the splash point. Each set of ripples proceeds to expand as though the other two ripple patterns were not there.

The principle of superposition states that every infinitisimal volume of the medium is capable of transmitting many discrete disturbances in many different directions, all simultaneously and with no detrimental effect of the one on the others. If we were able to observe and analyze the motion of a single air particle at a given instant under the influence of several disturbances we would find that it is the vector sum of the various particle motions required by all the disturbances passing by. At that instant that air particle

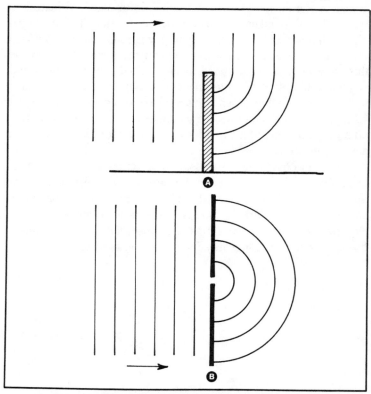

Fig. 4-8. (A) If the brick wall is large in terms of the wavelength of sound, the edge of the wall acts as a new source, radiating sound into the shadow zone. (B) Plane waves of sound impinging on a heavy plate with a small hole in it sets up spherical wave fronts on the other side due to diffraction of sound.

moves with amplitude and direction of vibration to satisfy the requirements of each disturbance just as a water particle responded to several disturbances in the ripple tank.

At a given point in space, let us assume an air particle responds to a passing disturbance with amplitude A and 0° direction. Another disturbance requires the same amplitude A but with a 180° direction. This air particle satisfies both disturbances at that instant by not moving at all.

ABSORPTION

The absorption of sound in the air of our great outdoors is a phenomenon involving viscosity, heat conductivity, and a relaxional behavior of the rotational energy states of the air, so complex that we gladly leave its consideration to the experts. Absorption of sound in fibrous or granular materials is primarily a matter of translation of sound energy to heat through friction. This frictional dissipation is usually provided by highly porous material in which the pores are open to each other. Common materials which provide this type of friction to sound include felted mineral fibers and vegetable fibers by the interstices between small granules, or a foam made of bubbles open to each other. When sound enters the surface of such a material, the air molecules vibrate at progressively reduced amplitude as friction is encountered. This friction acts as an acoustic resistance called the flow resistance and is measured by observing the pressure drop across a sample, as air of known velocity passes through the sample. This resistance must be within an optimum range: if too high there is little penetration and absorption suffers, if too low there is little absorption. An experienced acoustician might be observed blowing smoke through a sample to gauge its effectiveness as an absorber.

Acoustical Comb Filter Effects

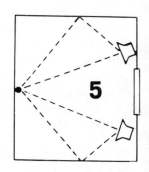

5

Phasing and flanging techniques are certainly well known today in the popular music field. Those using such effects generally associate them with the term "comb filter". Less widely appreciated is the fact that in many common recording and reproducing situations comb filter effects seriously distort our desired flat frequency response. These effects result in a form of amplitude distortion and they are inherent in practically every listening and monitoring setup, every single or multiple microphone mono pickup, as well as many mixdowns from stereo to mono. The magnitude of this distortion depends primarily upon the geometry of the setup, although other factors enter in.

Delay is the key word.[29] As an acoustical phenomenon, delay is a direct result of the finite speed of sound. For normal temperatures and elevations near sea level, sound travels 1,130 feet per second, or 1.13 feet per millisecond. In evaluating practical problems, a nice round figure to remember is that *sound travels about one foot per millisecond*.

A microphone is a rather blind sort of instrument. Its diaphragm responds to whatever fluctuations in air pressure occur at its surface. If the rate of such fluctuations (frequency) falls within its operating range it obliges with an output voltage proportional to the magnitude of the pressure involved. In Fig. 5-1 a 100 Hz tone from loudspeaker A actuates the diaphragm of a microphone in free space and a 100 Hz voltage appears at the microphone terminals. If a second loudspeaker B lays down a second 100 Hz signal at the diaphragm of the microphone identical in pressure, but 180° out of phase with the first signal, one acoustically cancels the other and the microphone voltage falls to zero. If an adjustment is made so that the two 100 Hz acoustical signals of identical amplitude are in phase, the microphone delivers twice the output voltage, an increase of 6 dB. The microphone slavishly responds to resultant pressures acting on its diaphragm. Little did it know (excuse the anthropomorphism) that when the two identical 100 Hz acoustical signals were in phase opposition that air molecules a short distance away from the diaphragm were obediently doing their violent 100 Hz dance. In short, the microphone responds to the vector sum of air pressure fluctuations impinging upon it. We must remember this characteristic of the microphone as we consider acoustical comb filter effects.

Now, as a slightly less than astounding revelation to those who are not sure just what a comb filter is, and as a review to those patient ones who do, we shall examine this ubiquitous effect in detail. We have seen that when two different airborne acoustical waves A and B of Fig. 5-1 arrive at a given point in space, such as our microphone diaphragm, they combine vectorially, that is, with due regard to both amplitude and phase. If A and B are identical sine waves of approximately the same amplitudes we have a highly simplified situation. With the 100 Hz example, combining in phase doubles the amplitude, combining in phase opposition results in cancellation. The same is true in combining identical, but highly complex speech and music signals.

It is helpful to consider how this interference effect acts down through the audio spectrum. Comb filter interference can radically affect the overall frequency response, even though each individual system component is flat. Let us assume that a microphone diaphragm is actuated by the combination of two signals, a signal direct from the mouth of one talking, and the same delayed 0.1 millisecond. Without the delayed signal let us say that the system response is flat and represented by the straight line at 0 dB in Fig. 5-2A. Adding to a given signal the same signal delayed 0.1 ms, the response undergoes some surprising changes. At those frequencies at which constructive interference takes place the response is boosted 6dB. Midway between the 6 dB peaks destructive interference creates dips infinitely deep, theoretically, 20 or 30 dB deep in practical situations. Significant energy is removed from the signal spectrum in the vicinity of 5 kHz and 15 kHz and unnatural

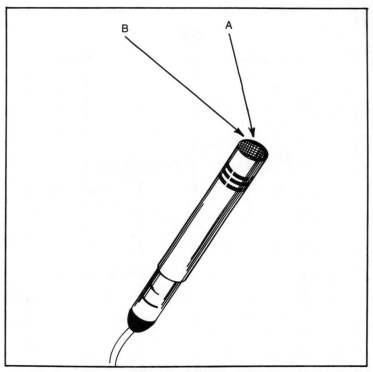

Fig. 5-1. The microphone diaphragm responds to the vector sum of sound pressures from multiple sources.

peaks introduced below 2 kHz, above 18 kHz, and in the 9-12 kHz region. Note that a linear frequency scale is used to show the symmetry of the peaks and dips.

If the delay is increased to 0.5 ms the peaks and dips are much closer together as shown in Fig. 5-2B. It is now apparent why the comb filter name was applied to this effect. Peaks now occur at 2, 4, 6, 8 kHz and at every other 2 kHz interval up through the spectrum. Between each pair of peaks is the accompanying dip. Figure 5-2C illustrates the comb filter effect when a signal is combined with itself delayed 1 ms. Peaks are now separated 1 kHz, as are the dips.

Now that it has served its purpose of showing the inherent symmetry of the comb filter response and the origin of the name, let us abandon the linear frequency scale for the more familiar logarithmic scale. Figure 5-3A shows the 0.1 ms delay case plotted in conventional semilog form. This gives a much better "feel" as to the effect of 0.1 ms delay on signal quality. The notches at 5 and 15 kHz would significantly color both speech and music. The 6 dB

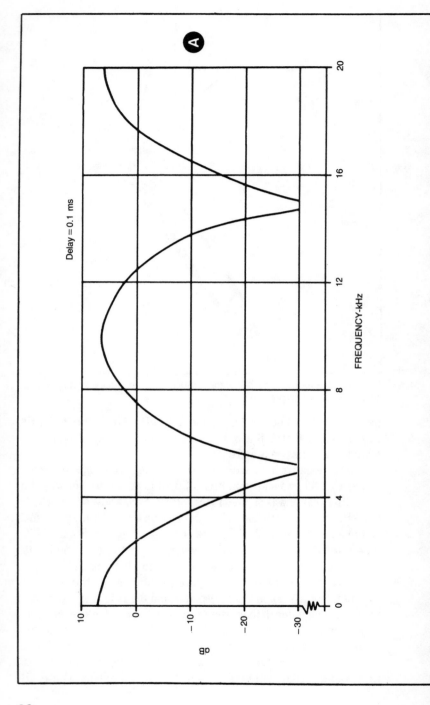

Delay = 0.1 ms

FREQUENCY-kHz

dB

Ⓐ

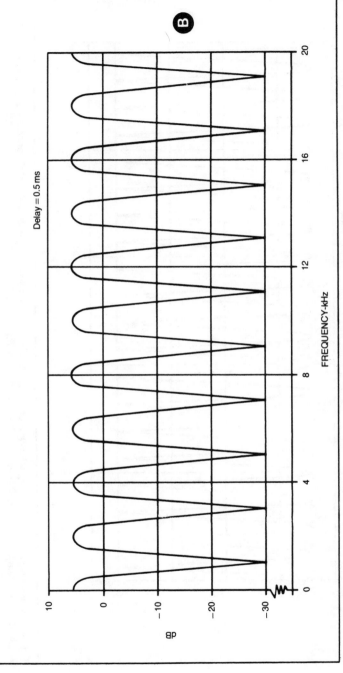

Fig. 5-2. Combining an acoustical signal with a delayed version of itself results in comb filter interference; (A) 0.1 ms delay, (B) 0.5 ms delay, (C) 1 ms delay. (Linear frequency scale).

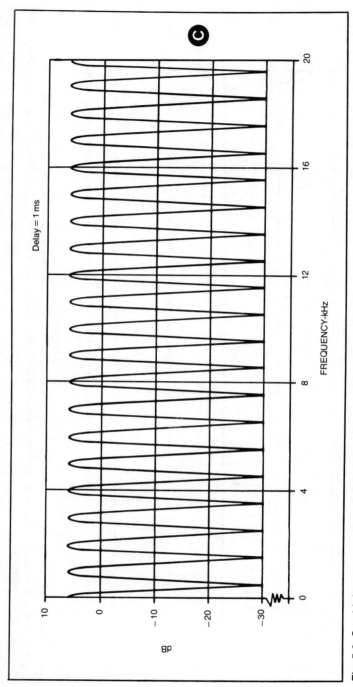

Fig. 5-2. Combining an acoustical signal with a delayed version of itself results in comb filter interference; (A) 0.1 ms delay, (B) 0.5 ms delay, (C) 1 ms delay. (Linear frequency scale). (Continued from page 81).

increase in level over significant portions of the spectrum may be viewed as a widening of the effective width of the dips.

The 0.5 ms delay case is presented in Fig. 5-3B on semilog coordinates. The dips appear very narrow in this plot, especially at the higher frequencies. Readers who have had experience with controlling feedback frequencies in sound reinforcement systems by applying numerous narrow notch filters might say that the dips of Fig. 5-3B might be tolerable, but not welcome. No matter how it is viewed, it is a significant deviation from a flat response.

Figure 5-3C illustrates the 1 ms delay example on a log frequency scale. Dips at 500, 1500, 2500 Hz, etc., are interspersed with peaks at 1, 2, 3 kHz, etc. Looking at all three parts of Fig. 5-3 we note the general principle that the longer the delay, the more the dips extend toward the low frequencies. Table 5-1 tabulates the location of the first null and spacings between adjacent nulls and adjacent peaks for delays from 0.1 ms to 50 ms. The same information in graphical form is shown in Fig. 5-4. The broken lines between the cancellation lines of Fig. 5-4 show, of course, the location of the peaks.

EFFECT OF RELATIVE AMPLITUDES

In Figs. 5-1, 5-2, and 5-3 it has been assumed that both the direct and the delayed signals are of equal amplitudes. In practical situations in which the delayed signal is the result of a less than perfect reflection, or arrives at the microphone at an angle at which response is down (e.g., off the axis of a cardioid mic), the delayed signal may be reduced in amplitude. Further, the inverse distance law hasn't been repealed. A component traveling farther arrives with lower amplitude than the component of signal traveling a direct path.

The boost will be less that 6 dB above normal and the nulls will be less than minus infinity if the delayed signal is less than the

Table 5-1. Comb Filter Peaks and Nulls.

Delay, ms	Frequency of lowest null, Hz	Spacing between nulls Spacing between peaks Hz
0.1	5,000	10,000
0.5	1,000	2,000
1.	500	1,000
5.	100	200
10.	50	100
50.	10	20

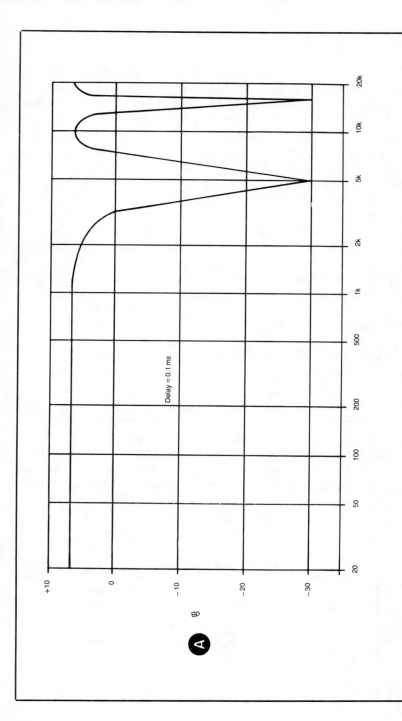

Delay = 0.1 ms

84

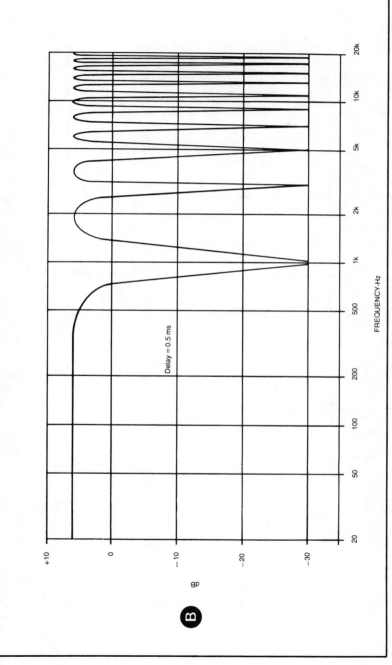

Fig. 5-3. The identical information of Fig. 5-2 plotted on the more familiar semilog coordinates; (A) 0.1 ms delay, (B) 0.5 ms delay, (C) 1 ms delay.

85

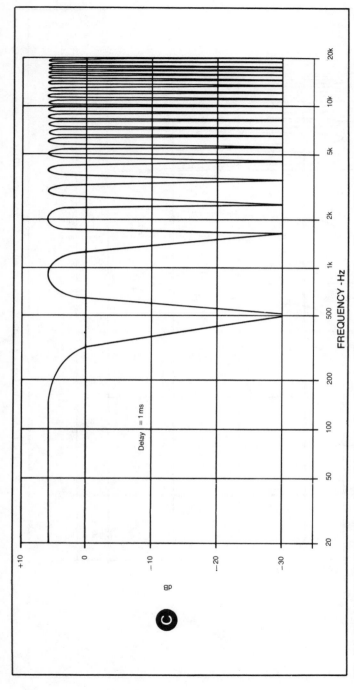

Fig. 5-3. The identical information of Fig. 5-2 plotted on the more familiar semilog coordinates; (A) 0.1 ms delay, (B) 0.5 ms delay, (C) 1 ms delay. (Continued from page 85).

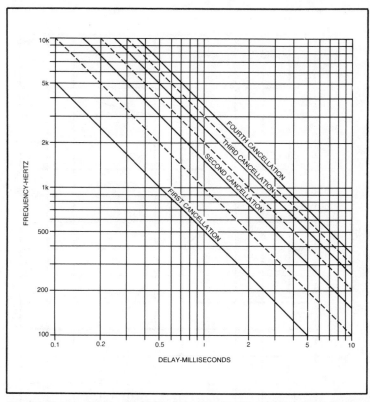

Fig. 5-4. The magnitude of the delay determines the frequency at which destructive interference (cancellations) and constructive interference (peaks) occur. The broken lines indicate the peaks between adjacent cancellations.

direct. Considering these two amplitudes as a ratio of unity or less, Fig. 5-5 enables one to determine theoretical peak height and null depth. If the delayed component is 80% of the direct, the peak height is about 5 dB and the null depth about 14 dB.

COMB FILTERS IN PRACTICE

There are many ways to generate comb filter effects. By having someone talk into a microphone as a hard wall is approached reveals quality changes due to comb filter interference. Another method giving better control is that of Fig. 5-6 in which a signal and a delayed version of itself are combined in a linear network. Applying a repetitive swept sine wave to the input and observing the output on a cathode ray oscilloscope for different delays, the responses of Figs. 5-2 and 5-3 are readily reproduced.

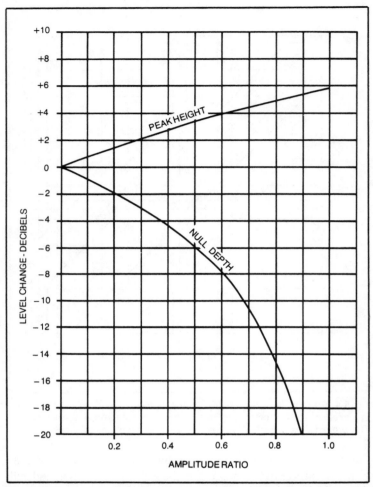

Fig. 5-5. The effect of amplitude ratios on comb filter peak height and null depth.

Example 1

Getting out of the laboratory and into the real world, consider the podium microphone arrangement of Fig. 5-7. Believe it or not, such arrangements with both mics feeding into the same amplifier can still be found. The excuse for using two mics this way is that it gives the talker greater freedom of movement, but how about interference effects? Assuming the microphones are properly polarized (and they might not be if this practice is typical of the local personnel), and if the talker is dead center, there would be a helpful level boost of 6 dB. What happens if he is 3 inches off the

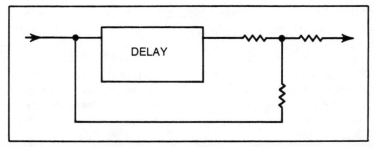

Fig. 5-6. The basic mechanism of comb filter generation is the same for both electrical and acoustical signals.

center line? Let us assume further that the microphones are 24 inches apart and that the talker's lips are 18 inches from a line drawn through the two microphones and on a level with the microphones. If the talker is centered, the sound travels the same distance (21.63″) to each microphone. If the talker moves laterally 3″, he comes closer to one microphone (20.12″) and increases his distance to the other (23.43″). The difference between these two distances (3.31″) results in the sound arriving at one microphone about 0.2 milliseconds behind the other. Result? A comb filter with a nice null gouging out important speech frequencies. If the talker were securely clamped in this position, the speech quality would not be good, but it would be unchanging.

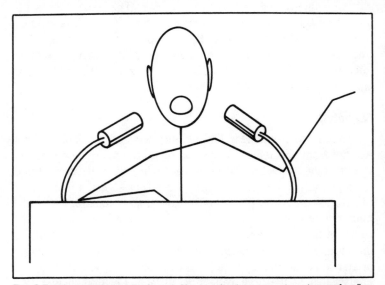

Fig. 5-7. Infamous example of comb filter production: two microphones feeding into the same mono amplifier.

Normal talker movements result in very noticeable changes in quality as the nulls and peaks shift up and down the frequency scale.

Welcome evidence that such effects are real and not just theoretical scare tactics has come out of the Electro-Voice anechoic chamber. Lou Burroughs gives numerous examples of wildly distorted responses due to what he calls "acoustic phase cancellation", or comb filter effects, measured in simulated setups.[30]

Example 2

Another situation much more common than Example 1 but similar in principle is the case of multiple sources close together, each source associated with its own microphone. Let us consider the singing group of Fig. 5-8 with each of the four singers with individual microphones. Singer A is d_1 inches from his or her own microphone and d_2 inches from the next microphone. The voice of A, picked up by both microphones, is mixed in the mixer with all the comb filter effects resulting from the path difference between d_1 and d_2. As each singer's voice is picked up, in one degree or another, by all microphones, the situation gets more and more complex. Fortunately, sounds picked up by the more distant microphones are weaker and the comb filter boosts and dips are correspondingly reduced as per Fig. 5-5. The experiments reported by Burroughs indicate that if singer A's mouth is at least three times farther from singer B's microphone than from his or her own, the comb filter effects are negligible. If all singers hold their microphones very close to their lips, the comb filter effects are submerged in other problems.

Example 3

Single microphones, like single men and women, are not exempt from problems. Concentrating on the microphone side of the analogy, reflecting surfaces result in sound arriving at the microphone somewhat later than the direct signal. Figure 5-9A illustrates the case of the talker or singer standing before a microphone mounted on a floor stand, or even hand held. In this case the floor reflected component (d_2) is much weaker than the direct (d_1) because (a) the distance of travel is greater, (b) the angle of arrival at the microphone is off the main axis, and (c) there is energy lost at the floor reflecting surface. Let us consider two simplified, specific cases, one in which $d_1 = 10''$ and another with d_1

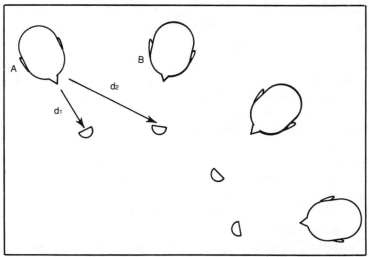

Fig. 5-8. For group singing, if d_2 is at least three times as great as d_1 the comb filter effect is minimized.

$= 50''$, both having a microphone height and soloist mouth height of $56''$, and a floor reflection coefficient of 0.95. We shall also assume a cardioid microphone which would give a response at 90° about 3 dB less than on axis. With the source 10" from the microphone, the floor reflected component is delayed about 7.6 ms. Considering only path length differences (10" direct vs. 112.4" reflected) we would expect the reflected component to be about 21 dB below the direct due to the inverse distance law). The reflection loss at the floor is only about 0.4 dB and the 90° off axis cardioid loss is another 3 dB. This places the reflected component more than 24 dB below the direct. Obviously, the resulting comb filter interference effect would be negligible.

The soloist now moves back to a point 50" from the microphone, either to improve "ambience", simply shrinking in fear, or for some other reason good enough for us to get on with this example. The direct path $d_1 = 50''$ is now somewhat more comparable to the reflected $d_2 = 122.7''$ and the reflected component arrives at the microphone 5.4 ms later than the direct. This would put the first null around 100 Hz, the second one near 300 Hz, which could be serious if the amplitudes are close enough in magnitude. It turns out that the inverse distance loss is about 7.8 dB to which we must add 0.4 dB for floor reflection and something like 1 dB for cardioid pattern 66° off axis for d_2 making a total of something like 9.2 dB. This corresponds to an amplitude ratio of about 0.35 which, from Fig. 5-5, indicates we can expect nulls about 4 dB deep and peaks about 2 dB high, or overall perturbations of our response of about 6 dB.

There are things we can do to improve the situation, if the soloist must be 50" from the microphone. A rug placed at the position of the floor bounce can be very effective in the upper audio range. A super or hyper-cardioid microphone could be used to reduce the reflected component another decibel or two. Thus our readily available remedies are quite limited, leaving the 50" distance to the microphone with basic problems unless a "shotgun" microphone is used which may well introduce a different set of problems.

Figure 5-9B shows a common geometry which can result in serious degradation of quality. Assume that $d_1 = 12''$ and $d_2 = 25''$. Then sound along the d_2 path would arrive close to 1 ms after d_1 which results in significant notches being taken from the signal spectrum. The amplitude ratio would be close to 0.5 and, referring to Fig. 5-5, we see that interference peaks would rise to about +3 dB and nulls would dip to about −6 dB giving overall response irregularities of about 9 dB. Closer talking will help this as well as a good sponge rubber pad on the desk top.

Distant microphone pickups, such as in Fig. 5-10A give distinct cancellation effects. By placing the microphone on the floor, as in Fig. 5-10B, d_1 is made approximately equal to d_2, thus minimizing the comb filter effect. There are several stands and foam protectors designed to support the microphone very close to the floor surface. Another recent approach is Crown International's Pressure Zone Microphone ™.[31]

Example 4

Radiating the same signal from two separate loudspeakers or groups of loudspeakers lays down a comb filter pattern over the audience area.[32] On the line of symmetry between the two groups of radiators signals arrive at the same time and no comb filter effects are noticed, at least this could be true for one ear with the other ear plugged. Moving to either side of this line means than the auditor is closer to one group than another and delays generate the classical comb filter effects. In areas where one loudspeaker is much stronger than the other due to path differences and resulting inverse distance fall off, the effects are modest. Directivity of the radiating sources also influence the area of the interference zone.

Reverberation will also influence the detectability of the effect. With loudspeakers 25 feet apart, the 5 ms contours are straight lines ± 12° from the line of symmetry down the center aisle.

Example 5

Multi-element loudspeakers can have their own private comb filter problems in the crossover region. In Fig. 5-11 it is apparent that frequency f_1 is radiated by both the bass and midrange units, that they are of essentially equal amplitudes, and that the two radiators are not physically at the same point. This means that at a

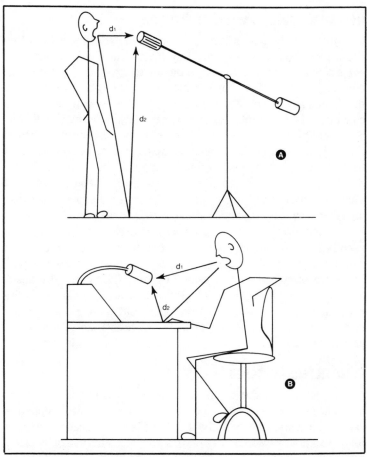

Fig. 5-9. Examples of the combination of direct and reflected components in common mono microphone setups in which the source is relatively close to the microphone and reflecting surface.

point in front of the loudspeaker the distance to the midrange unit may very well be different than the distance to the bass unit. This gives all the ingredients for generation of comb filter perturbations of response in the crossover region. The same process is active at f_2 between the midrange and tweeter units. Actually, a narrow band of frequencies is affected, the width limited by relative amplitudes of radiations from the two adjacent units. The steeper the crossover curves, the narrower the width of the frequency range affected. This is a highly complex problem that is being actively studied and loudspeakers claiming to minimize these and other time effects are appearing on the market.

HOW AUDIBLE ARE COMB FILTER EFFECTS?

Just how serious a threat to the quality of our signal is this comb filter business, anyway? Psychoacoustical research on the subject must eventually provide the answer to this question with any degree of finality. In the meantime subjective evaluations dominate. Aside from its use for special effects, one inescapable conclusion is that comb filter effects certainly color the signal - but how much? For one thing, this depends on the position and depth of the nulls which, in turn, depend on the magnitude of the delay and the amplitude of the delayed signal component as compared to the direct. In general, comb filter effects give a "roughness" and unnatural "edge" to the signal. There is no escaping the fact that signal energy lost in the nulls is lost forever to the ear.

Comb filter effects that are changing catch the attention more readily than unchanging ones. With two loudspeakers in the split system sound reinforcement, those seated in certain areas might be aware of considerable change in quality by moving the head. Walking down the aisle could add an undulating swishing to the program material radiated. In outdoor split loudspeaker setups refraction due to wind changes can introduce a variable swishing or a variable rough edge to the signal even to a stationary listener.

COMB FILTERS IN STEREO

In several of the comb filter examples considered earlier, mono has been specified. How about stereo? If two microphones are used in a stereo pickup and the sound is reproduced on a pair of loudspeakers, are comb filter effects present? There may be comb filter effects due to floor or other reflections in each of the two stereo microphones referred to above. The stereophonic information, however, is made up of the different amplitude and phase

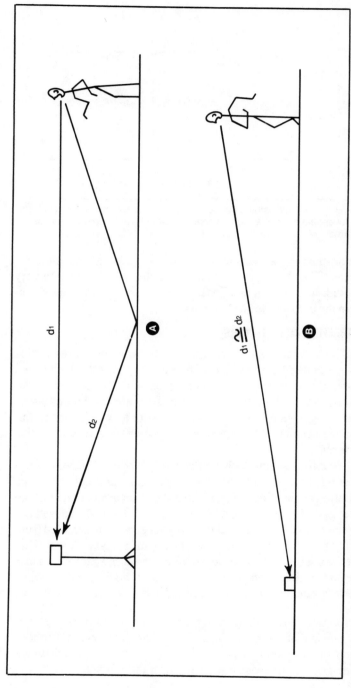

Fig. 5-10. Examples of the combination of direct and reflected components of sound in common mono microphone setups in which the source is at a distance from the microphone.

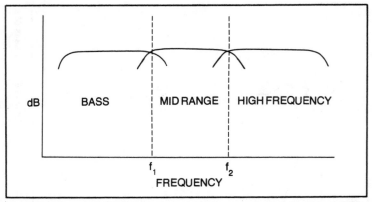

Fig. 5-11. Comb filter distortion can occur in the crossover region of a multiele-ment loudspeaker as the same signal is radiated from two units physically separated.

relationships as the two microphones pick up sounds from the same sources. The stereo image on playback depends upon these very amplitude differences and phase differences which are reassem-bled in the brain of the one listening.

ELIMINATING COMB FILTERS

The deleterious effects of comb filters, as discussed in this chapter, cannot be swept under the rug, they must be recognized and steps must be taken to overcome them if quality is our goal. An understanding of how comb filters are generated is the first step. Several suggestions have been offered as to how to minimize the peaks and nulls resulting from constructive and destructive interference.

Mounting the microphone flush with a table surface or other reflecting plane is an effective method of minimizing comb filtering for limited applications. Figure 5-12 shows an arrangement of the type suggested by Paul Veneklasen in the late sixties. A direct ray from the source S activates the microphone, but it is shielded from reflections from the surface in which the microphone is set. This minimizes alteration of response by comb filter action. Consider the effect of reflections from this surface if the microphone were pushed upward to position the active head a foot or so above that surface.

Another interesting thing about the flush mounted mi-crophone is that there is an increase in gain. The proximity of the microphone diaphragm to the reflecting surface of the table takes advantage of the increase in pressure near that surface.

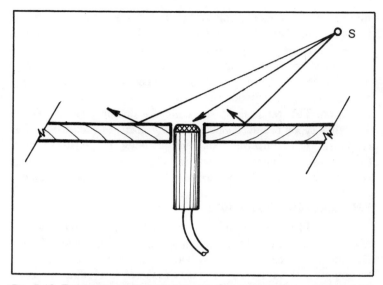

Fig. 5-12. The flush mounted microphone. Sounds from the source S which strike the surface do not reach the microphone and thus comb filter effects are avoided. Another advantage of this mounting is an increase of gain due to the pressure build up near the reflecting surface.

BASIC PROBLEMS OF MICROPHONES

Microphones are commonly calibrated in anechoic rooms or with special gating techniques so that the adverse effects of reflections are eliminated. However, these microphones are then used in rooms that are reverberant. The frequency response of the

Fig. 5-13. The Pressure Zone Microphone (see Ref. 133) which eliminates comb filter effects and is equally responsive to direct and reverberant components of the signal.

microphone responding only to the direct component is nice and flat when the microphone is oriented such that the sound is incident at zero degrees, i.e., parallel to the axis of the microphone and perpendicular to the plane of the diaphragm. How does this microphone respond to reflected sound coming from other directions? The high frequency response off axis falls off from the flat condition. This results in poor quality for distant pickups. There are other types of microphones (e.g., velocity) which have a flat response 90° from the axis and a rising response at high frequencies to sound off axis. In both cases the variation of response with angle of incidence results in coloration of the signal.

PRESSURE ZONE MICROPHONE

It would seem that one more microphone to add to the scores of types already available is something the world could well forgo. But the Pressure Zone Microphone[133] is a horse of a different wheelbase. It is based on the Pressure Recording Process conceived by Ed Long and Ron Wickersham in which the microphone element is placed very close to a reflective surface, a distance of 0.005 to 0.010 inch has been found to work best. As sound strikes a solid reflecting surface, particle velocity must be zero because the tiny air particles cannot move the massive surface. Zero particle velocity means maximum pressure and it is in this pressure zone that the microphone element is placed. The directional pattern of the microphone now becomes a hemisphere with essentially no discrimination between direct and reflected (reverberant) components, quite different from conventional microphones.

The Pressure Zone Microphone, shown in Fig. 5-13, consists of a metal plate and a supporting structure to hold the tiny microphone element with its diaphragm parallel to and only a few thousandths of an inch from the plate. The advantages are numerous. The omnidirectional pattern (hemispherical) has been mentioned. The frequency response is as flat as that of the basic element on-axis response except that low frequency response is affected by plate size. Typical response between −3 dB points is 50 Hz to 15 kHz for a small 2.5″ x 3″ plate. The response is the same to the direct sound field as to the reverberant sound field.

It is possible to use acoustical shields to alter the directional characteristic of the Pressure Zone Microphone. For example, a piece of foam can block response in a given direction, giving the microphone a unidirectional response.

The Pressure Zone Microphone is reported to yield a signal that is unusually clean and transparent. Moving sound sources, with their changing tonal quality resulting from shifting comb filter patterns in ordinary microphones, are clean with the Pressure Zone Microphone. Even though the war is not completely won, it does appear that a major weapon has been forged for the battle against comb filter effects.[134]

The Pressure Zone Microphone principle, however, has its critics. Stanley P. Lipshitz and John Vanderkooy of the University of Waterloo, Ontario, Canada, presented a paper before the Audio Engineering Society entitled, "The Acoustical Behavior of Pressure-Responding Microphones Positioned on Rigid Boundaries—A Review and Critique" (Preprint 1796-F-5, May 1981). These authors claim that a microphone mounted flush with a reflective surface, as in Fig. 5-12, shows excellent performance. Placing an "obstruction" (the PZM plate) between the diaphragm of the microphone and the incident sound degrades its performance. In order to regain acceptable performance a microphone having a much smaller diaphragm is used which results in greater self-noise. Such technical "give-and-take" is the stuff of which significant future improvements are made.

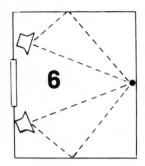

Sound
Indoors — á la mode

6

Sound in the great outdoors (Chapter 4) is subject to geometrical spreading, reflection, diffraction, superposition (interference), and absorption. But this is simple compared to what happens indoors. The antics of a roomful of air are almost beyond belief. In one degree or another all of the above listed effects are active indoors, and to these the added complication of resonance must be added, yet some understanding of these dominant resonance effects is vital to even a cursory appreciation of room acoustics. It is difficult to see how any task in the audio field can be carried out creatively and effectively without at least a basic understanding of how sound acts in enclosed spaces.

What shape of enclosure? To keep the complicated from getting even more complicated, the rectangular enclosure will be emphasized. After all, they are the most commonly used and the least expensive to build. Further, acoustical performance of rectangular enclosures has been studied far more than any other and a great body of important literature on this shape has been accumulated since the dawn of the 20th century. There has been no overwhelming evidence offered that other shapes are acoustically superior to the simple rectangular box. This chapter owes a great debt to researchers in the field of physical acoustics and names like Rayleigh, Sabine, Eyring, Knudsen, Morse, Hunt, Schroeder, Kuttruff, and Bolt add a strong foundation as well as great lustre to the field.

REFLECTIONS INDOORS

Anyone can observe a considerable difference between sound indoors and outdoors. Outdoors the only reflecting plane may be the earth's surface. If that surface happens to be covered with a foot of snow, which is an excellent absorber of sound, it may be difficult to carry on a conversation with someone 20 feet away. Indoors the sound energy is contained, resulting in a louder sound with a given effort and a speaker may be heard and understood by hundreds of people with no reinforcement but that of reflecting surfaces.

To enter this phase of our study we shall consider sound reflections from one wall at a time. In Fig. 6-1 a point source of sound, S, is a given distance from a massive wall. The spherical wave fronts (solid lines, traveling to the right) are reflected from this surface (broken lines). Physicists working in various forms of radiation (light, radio waves, sound) love to resort to the concept of images because it makes their mathematical studies much easier. In Fig. 6-1 the reflections from the surface traveling to the left act exactly as though they were radiated from another indentical point source, S_1, an equal distance from the reflecting surface but on the opposite side. This is the simplest image case of one source, one image, and a reflecting surface, all in free space.

Let us now bring the isolated reflecting surface of Fig. 6-1 down to earth and make it into the north wall of a rectangular room

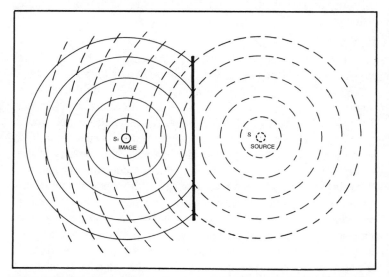

Fig. 6-1. Sound radiated by a point source S is reflected by the rigid wall. The reflected wave can be considered as coming from S₁, an image of S.

as in Fig. 6-2. The source S still has its image S_1 in what is now the north wall of the room. The source S also has other images. S_2 is the image in the east wall reflecting surface, S_3 is the image of S in the west wall and S_4 the image of S in the south wall. Use your imagination to visualize S_5, the image in the floor, and S_6, the image in the ceiling. All of these six images are assumed to be pulsating just like S, sending sound energy back into the room. Of course, the farther the images are from the reflecting plane, the weaker will be their contribution at a given point, P, in the room, but they all make their contribution.

We haven't seen the half of it. There are images of the images as well. The S_1 image has its image in the south wall at S_1', the image of the S_2 image in the west wall at S_2', and, similarly, images S_3, S_4, S_5, and S_6 images appear at S_3', S_4', and (not shown) S_5', and S_6'. I am reluctant to mention the images of the images of the images, but they are there, nonetheless, and so on *ad infinitum*. The more remote images are so weak that they may be neglected to make a given problem more tractable. Going further in discussion of the image is beyond the scope of this book. They are mentioned, to this extent, only to suggest a way to visualize how the sound field at some point P in a room is built up of the direct sound from the source S plus the vector sum of the contributions of all the images of S. This is only another way of saying that the sound at P is built up of the direct sound from S plus single or multiple reflections from all six surfaces.

TWO WALL RESONANCE

Figure 6-3 shows two parallel, reflective walls of infinite extent. If a loudspeaker radiating pink noise excited the space between the walls, the wall-air-wall system would exhibit a resonance at a frequency of $f_0 = 1130/2L$ or $565/L$, where L = the distance in feet between the two walls. A similar resonance would be set up at $2f_0$, $3f_0$, $4f_0$ . . . etc. down through the spectrum. The fundamental frequency f_0 is considered a natural frequency of the space between the reflective walls and it is accompanied by a train of modes which also exhibit resonances. Other names which have been applied to such resonances are *eigentones* (obsolete), *room resonances, permissable frequencies, natural frequencies,* or just plain *modes* which is preferred in this book. It takes more than two walls to make a studio or listening room. In adding two more pairs of walls, mutually perpendicular, to form a rectangular enclosure, we also add two more resonance systems, each with its own fundamental and modal series.

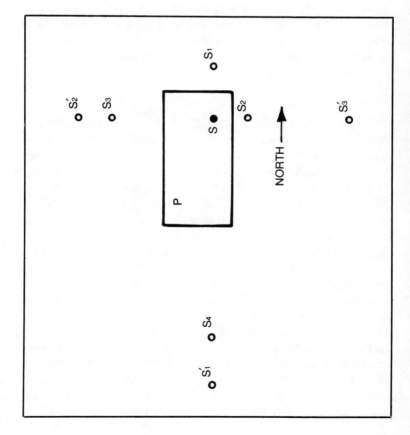

Fig. 6-2. The surface of Fig. 6-1 is made into the north wall of an enclosure. The source S now has six primary images, one in each of the six surfaces of the enclosure. Images of the images result in a theoretically infinite number of images of S. Sound intensity at point P is made up of the direct sound from S plus contributions from all images.

Actually, it gets far more complicated than that as Fig. 6-4 shows. So far we have been talking about *axial modes*, of which each rectangular room has three, plus modal trains for each. Each axial mode involves only two opposite and parallel surfaces. *Tangential modes*, on the other hand, involve four surfaces. *Oblique modes* go the full distance and involve all six surfaces of the room.

WAVES VS. RAYS

What the diagrams in Fig. 6-4 offer in terms of clarity, they lack in rigor. In these diagrams rays of sound are pictured as bouncing around on the angle of incidence equals the angle of reflection basis. For higher audio frequencies the ray concept is quite fruitful. When the size of the enclosure becomes comparable to the wavelength of the sound in it, however, special problems arise as the ray approach collapses. For example, a studio 30 feet long is only 1.3 wavelengths long at 50 Hz. Rays lose all meaning in such a case. Physicists employ the wave approach to study the behavior of sound in such an enclosure. This is an ongoing effort which continues to deepen our understanding of the performance of studios, control rooms, and other listening rooms, especially at the lower audio frequencies.

WAVE ACOUSTICS

At times like this it is very difficult to keep the treatments in this book non-mathematical. If we could dispense with such restraints, the first order of business would be to write down the wave equation. As a mere glance at this partial differential equation in three dimensions might strike terror to the hearts of many readers, such a temptation is resisted. Instead, we will drop down several steps from the basic wave equation to one of its solutions for sound in rectangular enclosures. The geometry used is that of Fig. 6-5 which judiciously fits the familiar mutually perpendicular x, y, z coordinates of three dimensional space to our studio or listening room. To satisfy orderly precepts, the longest dimension L is placed on the x axis, the next longest dimension W (for width) is placed on the y axis, and the smallest dimension H (for height) on the z axis. Who cares if we have to tip our room on its beam ends to do this? The sound inside will act the same. Our goal is to be able to calculate the permissable frequencies corresponding to the modes of a rectangular enclosure. Skipping the mathematics, as mentioned above, we come directly to the answer from the equation offered by Rayleigh in 1869:

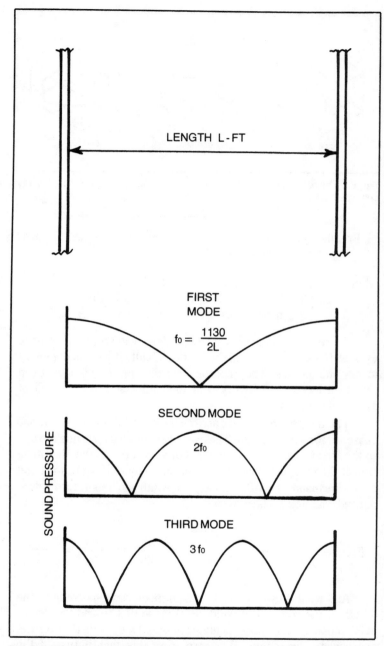

Fig. 6-3. Two parallel, reflective walls and the air between can be considered a resonant system with a frequency of resonance of $f_0 = 1130/2L$. This system is also resonant at integral multiples of f_0.

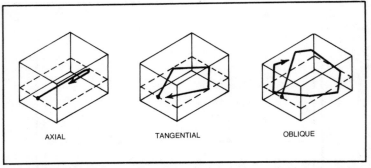

AXIAL   TANGENTIAL   OBLIQUE

Fig. 6-4. Visualization of axial, tangential, and oblique room modes by ray concept.

$$\text{Frequency} = \frac{c}{2} \sqrt{\frac{p^2}{L^2} + \frac{q^2}{W^2} + \frac{r^2}{H^2}} \qquad (6\text{-}1)$$

where,   c = speed of sound, 1130 ft/sec
           L, W, H = room length, width, and height, ft.
           p, q, r = integers 0, 1, 2, 3 . . . etc.

There are undoubtedly many readers who will flinch a little even at Equation 6-1. While it is not difficult, it can become a bit messy and tedious. The importance of this equation is that it can tell us the frequency of every axial, tangential, and oblique mode of a rectangular room.

The integers p, q, and r are the only variables once L, W, and H are set for a given room. These integers not only provide the key to the frequency of a given mode, but also serve as the identification of the mode as axial, tangential, or oblique. If p=1, q=0, and r=0 (shorthand, the 1,0,0 mode), the width and height terms drop out and equation 6-1 becomes:

$$\text{Frequency} = \frac{c}{2} \sqrt{\frac{p^2}{L^2}} = \frac{c}{2L} = \frac{1130}{2L} = \frac{565}{L}$$

And what do we have? The axial mode corresponding to the length of the room. The width axial mode (0,1,0) and the height axial mode (0,0,1) are calculated similarly by substituting the appropriate dimension. We learn from this that if two of the integers are zero, an axial mode frequency is identified, because only one pair of surfaces is involved. In a similar way we learn that one zero identifies tangential modes, and no zero at all specifies oblique modes.

MODE CALCULATIONS — AN EXAMPLE

The utility of Equation 6-1 is best demonstrated by a specific example. The room is small, but convenient for experimental verification. The dimensions of this room are:
length L = 12.46 feet, width W = 11.42 feet, and average height H = 7.90 feet (the ceiling actually slopes along the length of the room with a height of 7.13 feet on one end and 8.67 feet on the other). These values of L, W, and H have been inserted in Equation 6-1 along with an assortment of combinations of integers p, q, and r.

Room modes are possible only when p, q, and r are whole numbers (or zero) as this is the condition that will create a *standing* wave. There are many combinations of integers when fundamentals (associated with 1), second modes (associated with 2), third mode (associated with 3), etc., are introduced. Table 6-1 lists some of the various combinations of p, q, and r and the resulting permissible modal frequency for each combination. Further, each frequency is identified as axial, tangential, or oblique by the number of zeros in that particular p, q, r combination. The lowest

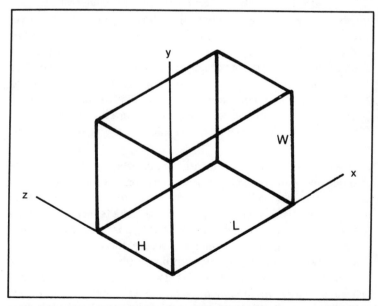

Fig. 6-5. Orientation of rectangular room of length L, width W, and height H with respect to the x, y, and z coordinates for calculation of room modal frequencies.

natural room frequency is 45.3 Hz which is the axial mode associated with the longest dimension, the length L, of the room. In this case p = 1, q = 0, and r = 0. Mode 7, the 2,0,0 mode, yields a frequency of 90.7 Hz which is the second axial mode associated with the length L. In the same manner mode 18, with integers 3,0,0, is the third mode of the length axial mode, and mode 34 the fourth mode. But there are many tangential and oblique modes between these and Table 6-1 provides the means of carefully identifying all modes and raises the possibility of studying their relationships.

The axial modes have been emphasized in studio design and the good reason for doing this will be discussed later. The message of Equation 6-1 and Table 6-1 is that there is much more to room acoustics than axial modes and their spacing. In between axial modal frequencies are many other modal frequencies which have an effect, even though weaker.

EXPERIMENTAL VERIFICATION

All of the modal frequencies listed in Table 6-1 make up the acoustics of this particular room for the frequency range specified. To evaluate their relative effects a swept sine wave transmission experiment was set up. Knowing that all room modes terminate in the corners of a room, a loudspeaker was placed in one low corner and a measuring microphone in the diagonal high corner of the room. The loudspeaker was then energized by a slowly swept sine wave signal. The room response to this signal, picked up by the microphone, was recorded on a graphic level recorder having a paper speed of 3 mm/second. This resulted in a linear sweep from 50 to 250 Hz in 38 seconds. The resulting trace is shown in Fig. 6-6.

Attempts in the past to identify the effects of each mode in this manner have been made in reverberation rooms with all six surfaces hard and reflective. In such cases the prominent modes stand out as sharp spikes on the recording. The test room in which the recording of Fig. 6-6 was made is a spare bedroom, not a reverberation chamber. Instead of concrete, the walls are of frame construction covered with gypsum board (drywall); carpet over plywood make up the floor, closet doors almost cover one wall. There is a large window, pictures on the wall, and some furniture including a couch. It is evident that this is a fairly absorbent room. The reverberation time at 125 Hz (as we shall consider more fully

Table 6-1. Mode Calculations.

Mode Number	Integers p q r	Mode Frequency, Hz	Axial	Tangential	Oblique
1	1 0 0	45.3	x		
2	0 1 0	49.5	x		
3	1 1 0	67.1		x	
4	0 0 1	71.5	x		
5	1 0 1	84.7		x	
6	0 1 1	87.0		x	
7	2 0 0	90.7	x		
8	2 0 1	90.7		x	
9	1 1 1	98.1			x
10	0 2 0	98.9	x		
11	2 1 0	103.3		x	
12	1 2 0	108.8		x	
13	0 2 1	122.1		x	
14	0 1 2	122.1		x	
15	2 1 1	125.6			x
16	1 2 1	130.2			x
17	2 2 0	134.2		x	
18	3 0 0	136.0	x		
19	0 0 2	143.0	x		
20	3 1 0	144.8		x	
21	0 3 0	148.4	x		
22	2 2 1	152.1			x
23	3 0 1	153.7		x	
24	1 1 2	158.0			x
25	3 1 1	161.5			x
26	0 3 1	164.8		x	
27	3 2 0	168.2		x	
28	2 0 2	169.4		x	
29	1 3 1	170.9			x
30	0 2 2	173.9		x	
31	2 3 0	173.9		x	
32	2 1 2	176.4			x
33	1 2 2	179.7			x
34	4 0 0	181.4	x		
35	3 2 1	182.8			x
36	2 3 1	188.1			x
37	2 2 2	196.2			x
38	0 4 0	197.9	x		
39	3 0 2	197.9		x	
40	3 3 0	201.3		x	
41	3 1 2	203.5			x
42	0 3 2	206.1		x	
43	1 3 2	211.1			x
44	0 0 3	214.6	x		
45	1 0 3	219.3		x	
46	0 1 3	220.2			x
47	3 2 2	220.8			x
48	1 1 3	224.8			x
49	2 3 2	225.2			x
50	2 0 3	232.9		x	

Table 6-1. Mode Calculations (continued from page 109).

Mode Number	Integers p q r	Mode Frequency	Axial	Tangential	Oblique
51	4 3 0	234.4		x	
52	0 2 3	236.3		x	
53	2 1 3	238.1		x	
54	3 4 0	240.2		x	
55	1 2 3	240.6			x
56	3 3 2	247.0			x
57	2 2 3	253.1			x
58	3 0 3	254.0		x	
59	0 3 3	260.9		x	
60	3 2 3	272.6			x
61	2 3 3	276.2			x
62	4 0 3	281.0		x	
63	0 0 4	286.1	x		
64	0 4 3	291.1		x	
65	3 0 4	316.8		x	
66	0 3 4	322.3		x	

later) was found to be 0.33 second. This room is acoustically much closer to studios and control rooms than to reverberation chambers, and this is why it was chosen.

MODE IDENTIFICATION

A careful study of Fig. 6-6 in an attempt to tie the peaks and valleys of the transmission run to specific axial, tangential, and oblique modes is rather disappointing. For one thing, the loudspeaker(JBL 2135) response is included, although it is quite smooth over this frequency region. The signal generator and power amplifier are very flat. Hence, we must attribute most of these ups and downs to modes and the interaction of modes. Modes close together would be expected to boost room response if in phase, but cancel if out of phase. There are 11 axial, 26 tangential, and 21 oblique modes in this 45.3 -254.0 Hz record and the best we can say is that the transmission trace shown in Fig. 6-6 is the composite effect of all 58 modes.

The three major dips are so narrow that they would take little energy from a distributed speech or music spectrum. If they are neglected the remaining fluctuations are more modest. Fluctuations of this magnitude in such steady state swept sine transmission tests are characteristic of even the most carefully designed and most pleasing of studios, control rooms, and listening rooms. Our ear commonly accepts such deviations from flatness of

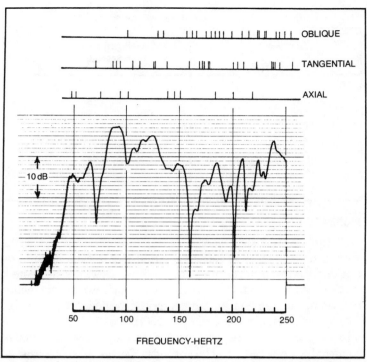

Fig. 6-6. Swept sine wave transmission run in the test room of Table 6-1. The location of every axial, tangential, and oblique modal frequency is indicated.

response. The modal structure of a space always gives rise to such fluctuations. They are normally neglected as attention is focussed on the response of microphone, amplifier, loudspeaker, and other reproducing equipment. The response of the ear and of the room have yet to receive the attention they deserve.

Table 6-2. Modes in Octave Bands.

Octave	Limits (−3 dB points)	Modes		
		Axial	Tangential	Oblique
63 Hz	45-89 Hz	3	3	0
125	89-177 Hz	5	13	7
250	177-354	3*	10*	14*
		*For the range 177-254 Hz		

Table 6-3. Measured Reverberation Time of Test Room.

Frequency Hertz	Average Reverberation Time, Seconds
180	0.38
200	0.48
210	0.53
220	0.55
240	0.31 and 0.53 (double slope)
125 Hz octave noise	0.33
250 Hz octave noise	0.37

MODE DECAY

The steady state response of Fig. 6-6 tells only part of the story and the ear is very sensitive to transient effects. Reverberation decay is one transient phenomenon in which we are interested. When broad band sound such as speech or music excites the modes of a room, our interest naturally centers on the decay of the modes. The 58 varieties of modes in the 50 - 250Hz band of Fig. 6-6 are the microstructure of the reverberation of the room. Reverberation is commonly measured in octave bands. Octave band widths of interest range as shown in Table 6-2.

Each reverberatory decay by octave bands thus involves an average of the decay of many modes. The higher the frequency of the band, the more modes included.

The question arises, do all modes decay at the same rate? The answer is a resounding, "No!". It depends, among other things, on how absorbing material is distributed in the room. Carpet on the floor of the test room has no effect whatever on the 1,0,0 or 0,1,0 axial modes involving only walls. Further, tangential and oblique modes, involving more surfaces are expected to die away faster than axial modes involving only two surfaces.

So much for expectation. As the detective says, "Just give me the facts, lady". Actual reverberation decays in the test room using sine waves are shown in Figs. 6-7 and 6-8. The measured reverberation times varied as is shown in Table 6-3.

We find that measured reverberation time varies almost two to one for the selected frequencies in the 180 - 240 Hz range. The decay of the 125 Hz octave band of noise (0.33 sec.) and the 250 Hz octave band (0.37 sec.) of Fig. 6-8 averages many modes and should be considered more or less the "true" values for this frequency region, although normally many decays for each band would be taken to provide statistical significance.

The dual slope decay at 240 Hz is especially interesting as the low value of the early slope (0.31 sec) is probably dominated by a

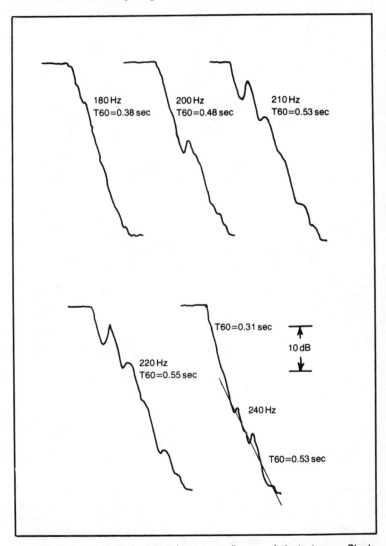

Fig. 6-7. Pure tone reverberation decay recordings made in test room. Single modes decaying alone yield smooth, logarithmic trace. Beats between neighboring modes cause the irregular decay. The two-slope pattern (bottom, right) reveals a smooth decay of a single prominent mode for the first 20 dB below which one or more lightly absorbed modes dominate.

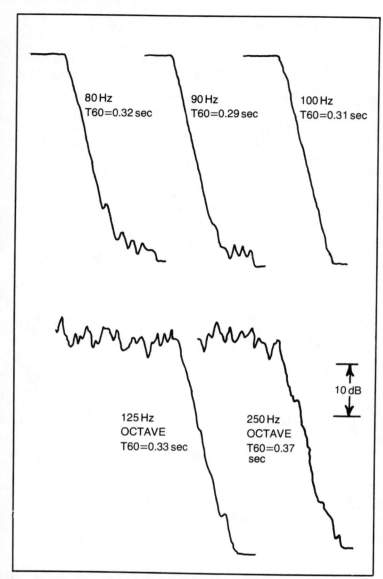

Fig. 6-8. The top three pure tone reverberation decays are dominated by single, prominent modes as indicated by the smooth adherence to logarithmic form. The two lower records are of octave bands of pink noise which give the average decay of all modes in those octaves.

single mode involving much absorption, later giving way to other modes which encounter much less absorption. Actually identifying the modes from Table 6-1 is difficult although one might expect mode 44 to die away slowly and the group of three near 220 Hz (45, 46, 47) to be highly damped. It is common to force nearby modes into oscillation which then decay at their natural frequency.

MODE BANDWIDTH

Normal modes were introduced as part and parcel of room resonances. Taken separately, each normal mode exhibits a resonance curve such as shown in Fig. 6-9. Each mode, therefore, has a definite bandwidth determined by the simple expression:

$$\text{Bandwidth} = f_2 - f_1 = \frac{2.2}{\text{RT60}} \tag{6-2}$$

RT60 = reverberation time, seconds

Bandwidth is inversely proportional to the reverberation time. In electrical circuits we recall that the sharpness of the tuning curve depends on the amount of resistance in the circuit, the greater the resistance the broader the tuning curve. In room acoustics the reverberation time depends on absorption (resis-

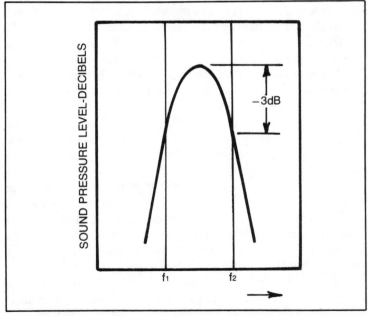

Fig. 6-9. Each mode has a bandwidth. The more absorbent the room, the greater the bandwidth. As measured at the customary −3 dB points, the bandwidths of a recording studio may be about 5 Hz.

115

tance) and the analogy is fitting (the more absorption, the shorter the reverberation time, the wider the mode resonance). For convenient reference, Table 6-4 lists a few values of bandwidth in relation to reverberation time.

The mode bandwidth for most studios is in the general region of 5 Hz. This means that for rooms of low reverberation time adjacent modes tend to overlap, which is desirable.

An expanded look at the 40 - 100 Hz portion of Fig. 6-6 is shown in Fig. 6-10. The axial modes from Table 6-1 falling within this range are plotted with bandwidth of 6 Hz at the -3dB points. The axial mode peaks, Fig. 6-10A, are taken as zero reference level. The tangential modes have only ½ the energy of the axial modes[33] which justifies plotting their peaks 3 dB below the axial modes in Fig. 6-10B. The oblique modes have only ¼th the energy of the axial modes, hence the lone oblique mode at 98.1 Hz falling within this frequency range is plotted 6dB below the axial mode peaks in Fig. 6-10C.

The response of the test room, Fig. 6-10D, is most certainly made up of the collective contributions of the various modes tabulated in Table 6-1. Can D of Fig. 6-10 be accounted for by the collective contributions of axial modes A, tangential modes B, and the single oblique mode C? It seems reasonable to account for the 12 dB peak in the room response between 80 and 100 Hz by the combined effect of the 2 axial, 3 tangential, and 1 oblique mode in that frequency range. The fall off below 50 Hz is undoubtedly due to loudspeaker response. This leaves the 12 dB dip at 74 Hz as a main feature of the response to be accounted for.

Examining the axial mode at 71.5 Hz we are reminded that it is the vertical mode of the test room working against a sloping ceiling. The frequency corresponding to the average height is 71.5 Hz, but that corresponding to the height at the low end of the ceiling is 79.3 Hz and that for the high end is 65.2 Hz. The uncertainty of the frequency of this mode is indicated by arrows in Fig. 6-10. If this uncertain axial mode were shifted to a slightly lower frequency, the 12dB dip in response could be better explained. It would seem that a dip in response should have been recorded near 60Hz, but none was found.

Rather than being an absolute experimental verification of theory, this test room experiment was conducted only to explain that theory. Conditions lacked the rigor to expect exact results. The test room is not a rectangular parallelepiped. The shape of the loudspeaker response is known only approximately, plus other

Table 6-4. Mode Bandwidth.

Reverberation Time, seconds	Mode Bandwidth hertz
0.2	11
0.3	7
0.4	5.5
0.5	4.4
0.8	2.7
1.0	2.2

minor uncertainties. Figure 6-10 emphasizes the main thrust of this chapter, that room acoustics is determined by room modes.

MODE SUMMARY

■ There are three types of acoustical resonances (natural frequencies, standing waves, normal modes) in a rectangular enclosure; axial, tangential, and oblique modes.

■ The axial modes are made up of two waves going in opposite directions, traveling parallel to one axis, and striking only two walls. Axial modes make the most prominent contribution to the acoustical characteristics of a space. As there are three axes to a rectangular room, there are three fundamental axial frequencies, each with its train of modes.

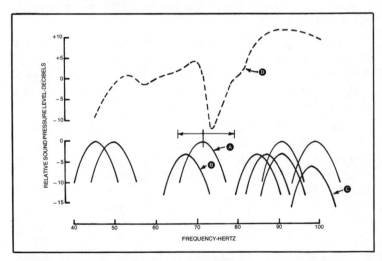

Fig. 6-10. Attempted correlation of calculated modes and the measured swept sine response of the test room over the frequency range 40 to 100 Hz; (A) axial modes, (B) tangential modes, (C) oblique mode, (D) reproduction of the 40 to 100 Hz portion of room response of Fig. 6-7.

117

■ The tangential modes are formed by four traveling waves reflecting from four walls and moving parallel to two walls. Tangential modes have only half the energy of axial modes, yet their effect on room acoustics may be significant. Each tangential mode has its train of modes.

■ The oblique modes involve eight traveling waves reflecting from all six walls of an enclosure. Oblique modes, having only ¼th the energy of the axial modes, are of less effect than the other two.

■ The number of normal modes increases with frequency. Small rooms, whose dimensions are comparable to the wavelength of audible sound, have the problem of the "piling up" of modes or excessive separation which can contribute to poor characteristics for recording or other critical work. A musical note falling "in the cracks" between widely separated modes, will be abnormally weak and will die away faster than other notes. It is almost as though that particular note were sounded outdoors while the other notes were simultaneously sounded indoors.

■ Axial, tangential, and oblique modes decay at different rates. Absorbing material must be located on surfaces near which a given modal pressure is high if it is to be effective in absorbing that mode. For example, carpet on the floor has no effect on the horizontal axial modes. Tangential modes are associated with more surface reflections than axial modes and oblique modes even more than tangential.

■ As frequency is increased, the number of modes greatly increases. Above 300Hz average mode spacing becomes so small that room response tends to become smoother. Greater energy is contained in oblique and tangential modes at high frequencies because of their great number.

■ Colorations caused by acoustical anomalies of studios, monitoring rooms, listening rooms, and other small rooms are particularly devastating to speech quality. Gilford states[34] that axial modes spaced approximately 20 Hz or more, or a pair of modes coincident or very close, are frequent sources of colorations. He also states that colorations are likely to be audible when an axial mode coincides with a fundamental or first formant of at least one vowel sound of speech and are in the region of high speech energy. Speech colorations below 80Hz are rare because so little speech energy is in that part of the spectrum. There are essentially no speech colorations above 300Hz for either male or female voices.

■ The bandwidth of room modes, measured at the −3dB points, increases with a shortening of reverberation time. Modes in ordinary studios have bandwidths in the order of 5 Hz. Harmonics of modes have the same bandwidths as their fundamentals.

For those who wish to pursue this subject further the reader is urged to consult references 35 through 46. These papers, spanning a half century, trace the growth in understanding of room modes.

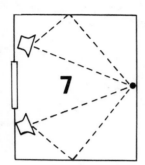

7 Echoes in Smaller Rooms

An *echo* is usually defined as a reflected wave of sufficient amplitude and delay to make it distinct from the sound giving rise to it. A yodel from an alp may be reflected from the face of another distant alp, producing a weak repeat of the outgoing sound which may be considered quaint and interesting. In concert halls and large auditoriums, faulty acoustical design may result in large reflecting surfaces which may be considered a distinct pain in the acoustic cortex because of the echoes they create. An important part of the practice of architectural acoustics is minimizing echo production.

In this chapter interest is focused on echoes in radio, television, and recording studios, monitoring and other listening rooms. Impulsive sounds reflected from various surfaces may be observed on the screen of a cathode ray oscilloscope as discrete packets of energy, but the ear does not perceive them as full blown, discrete echoes with short delays characterizing the smaller enclosures. Instead, there may be an incipient echo effect, an awareness that something echolike is there, yet not in discrete form. Such embryonic echo effects often show up as a result of acoustical flaws such as flat reflecting surfaces, corner reflections, etc. Proper application of diffusing techniques can often ameliorate the incipient echo effect.

Rooms have a different "sound" than outdoors due to the containment of sound energy offered by the room surfaces. A given effort produces a louder sound indoors and, if conditions are right, we like the indoor effect. Colorations of the original sound,

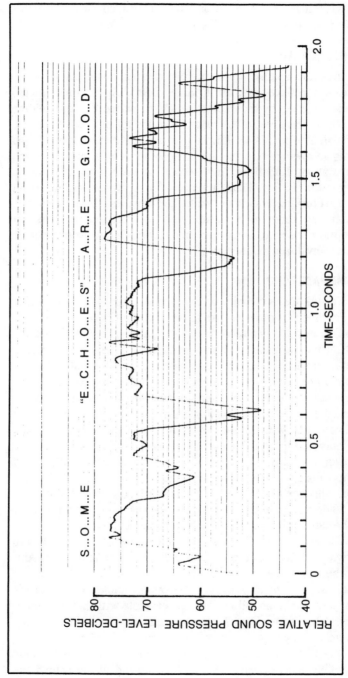

Fig. 7-1. High speed graphic level recording of the phrase, "Some echoes are good". Made with Brüel & Kjaer 2305 graphic level recorder, writing speed 1,000 mm/sec, paper speed 100 mm/sec.

however, must be avoided and the echoes considered in this chapter are one of the several sources of colorations.

MEAN FREE PATH

In small rooms sound travels shorter distances between successive reflections than in large rooms. An average (mean) distance is given by the expression 4V/S, where V is the volume and S the surface area of the space. In an auditorium 125 x 65 x 45 ft the mean free path is 43.9 ft. In a studio 23 x 19 x 15 ft the sound travels an average of only 12.3 ft. between reflections. As sound travels about 1.13 ft per millisecond, it takes only 10.9 ms to traverse this 12.3 ft. Thus, in terms of time, we can expect sound to start returning from floor, walls, and ceiling surfaces of a studio or listening room in a matter of a few milliseconds. In our perception of sound, some very interesting things happen during the first few dozen milliseconds.

A HOMEMADE ECHO

Echoes are as complex as the original sound which gives rise to them, hence it makes sense to concentrate first on the original sound. In Fig. 7-1 is a recording of the phrase, "Some echoes are good". This was made on a high speed graphic level recorder with a writing speed of 1,000 mm per second to follow the rapid changes of sound pressure. Let us assume that this train of sound waves travels out and is reflected from a distant surface. The echo that is returned, for the purposes of this illustration, has exactly the same level variations with time as the original, but it is of lower level and it is delayed. Figure 7-2 illustrates the case in which the echo is 28 dB lower in level than the original and is delayed 205 ms. A delay of 205 ms tells us that the sound traveled a distance of (205) (1.13) = 232 ft. This could be an echo from a reflective surface 116 ft away for a 90 degree reflection or the difference between the direct path and the path taken by the reflected component.

The echo in Fig. 7-2 has been shifted downward 28 dB and to the right 205 ms until the original and echo touch at point A. Any time the echo level momentarily attains the level of the original sound, such as at point A, the echo becomes audible. At least this is the hypothesis being followed in this example with justification to follow. Therefore, Fig. 7-2 represents one condition of echo level-echo delay in which the echo is audible and, of course, there are many others. Figure 7-3 is a plot of echo level vs. echo delay for just audible echo and our — 28 dB echo level and 205 ms echo delay

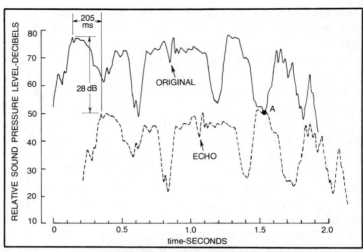

Fig. 7-2. An echo of the signal of Fig. 7-1 is simulated by displacing a tracing of it downward (for lowering its level) and to the right (to introduce a time delay).

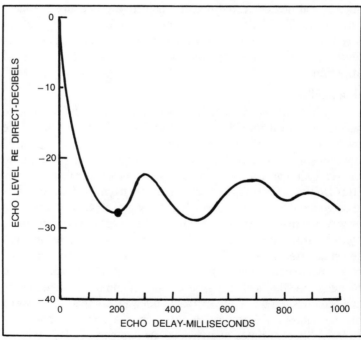

Fig. 7-3. A correlogram derived from the signal recording of Fig. 7-1 by the method illustrated in Fig. 7-2. Each combination of echo level and echo delay resulting in echo level being equal to the level of the original yields one point on this graph. The heavy spot is point A, derived from Fig. 7-2.

123

condition is shown as a black dot. The rest of the graph of Fig. 7-3 was obtained by shifting the echo trace horizontally and vertically until the two traces touch and plotting the delay and level pertaining to each such condition. Figure 7-3 tells us that echoes with short delays must be much stronger to be heard than echoes with longer delays, but the graph tends to level off for greatly delayed echoes.

The graph of Fig. 7-3 is called a correlogram. It is only as accurate in determining the audibility of echoes of the "Some echoes are good" phrase as the assumptions underlying the procedure. These assumptions are, (1) that the spectrum of the echo is identical to that of the original sound, (2) that directional effects are negligible, (3) that the ear's decay rate is the same as that of the graphic level recorder (in our case, 500 dB/sec or 120 ms for 60 dB decay), (4) that an echo is perceptible if its level is equal to or greater than the level of the signal during any moment of their simultaneous presentation. It turns out that none of these are really too far off base. Dubout[47] ran many subjective tests for speech and demonstrated that the objective procedure we have followed in Figs. 7-1 through 7-3 is verified by his listening tests to a remarkable degree, at least for echoes of greater than 50 ms delay. Echoes in this shorter range will now receive special attention.

HAAS EFFECT

Helmut Haas didn't exactly discover the "Haas Effect" but he surely brought it forcibly to our attention with his research[48] He made a significant statement when he wrote: "Our hearing mechanism integrates the sound intensities over short time intervals similar, as it were, to a ballistic measuring instrument". Herr Doktor Haas set his subjects 3 meters from two loudspeakers arranged so that they subtended an angle of 45 degrees, the observer's line of symmetry splitting this angle. The conditions were approximately anechoic. The observers were called upon to adjust an attenuator until the sound from the "direct" loudspeaker was equal to that of the "delayed" loudspeaker. He then proceeded to study the effects of varying the delay. A number of researchers had previously found that very short delays (less than 1 ms) were involved in our discerning the direction to a source of sound by slightly different times of arrival at our two ears. Delays greater than this do not affect our directional sense.

As shown in Fig. 7-4 (a repeat of Fig. 2-14 for convenience) Haas found that in the 5-35 ms delay region the sound from the delayed loudspeaker had to be increased more than 10 dB over the direct before it sounded like an echo. This is the precedence or Haas Effect. In a small room such as a studio or listening room reflected energy arriving at the ear within 50 ms is integrated with the direct and is perceived as part of the direct sound as opposed to reverberant sound. These early reflections increase the loudness of the sound and, as Haas has said, result in "a pleasant modification of the sound impression in the sense of broadening of the primary sound source while the echo source is not perceived acoustically".

The transition zone between the integrating effect, for delays less than 50 ms, and the perception of delayed sound as a discrete echo is gradual and therefore somewhat indefinite. Some place the dividing line at 1/16th second (62 ms), some at 80 ms, and some at 100 ms beyond which there is no question of the discreteness of the echo. For the purposes of this discussion we shall arbitrarily consider the first 50 ms, as in Fig. 7-4, the region of definite integration.

INTERMEDIATE DELAYS

Haas' work, published in 1951, precipitated further investigations in many countries. Selections from the results of one of

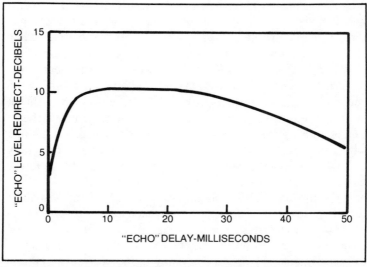

Fig. 7-4. The precedence zone as determined by Haas (Ref. 48). This curve describes the amount the delayed signal must be increased in level to sound as loud as the undelayed, direct sound. The delayed signal is perceived as an echo only if it exceeds this value.

these[49] is shown in Fig. 7-5. These tests were conducted in spaces having reverberation times of 0.5 and 0.2 second, comparable to some recording studios, control rooms, and even many living rooms. Only echo delays less than 120 ms are shown as they best apply to studios and other smaller rooms in contrast to large auditoriums. Curves for 20% and 50% of observers being disturbed by the echo are shown in Fig. 7-5. This practice results from the experimental problems in obtaining consistent results for reporting the onset of echo. Note the general similarity in shape to the crude curve of Fig. 7-3 which we derived from Figs. 7-1 and 7-2 except that the method used in deriving Fig. 7-3 does not incorporate the important integration effect for echoes arriving within 50 ms.

For short echo delays the curves of Fig. 7-5 go above the zero level line which means that the echo level is higher than that of the direct sound and that the integrating effect is present for these conditions. Because the conditions of this experiment are different from those of Haas' experiments, the exact shape of the curves differ somewhat from Fig. 7-4.

An exponential decay is a straight line on a graph of dB vs. time such as Fig. 7-5. A reverberation time of 0.5 second means that sound decays 60 dB in 0.5 second. A slight exercise in proportions tells us that sound would decay 10 dB in (10/60) (0.5) = 0.083 second. Plotting this point and drawing a straight line through it and through zero gives us the decay rate corresponding to RT60 = 0.5 second. It is interesting to note that the broken line of Fig. 7-5 for RT60 = 0.5 second is more or less tangent to the 20% curve obtained in a space having a reverberation time of 0.5 second. This makes sense as we recall (1) that the 20% curve is not far from the threshold of echo perception, and (2) that reverberation is made up of all the echoes as they are reduced in amplitude with time. Apart from the short delay integration region, this really confirms the idea that when echoes are equal in amplitude to the direct signal, they become audible.

For the lower curves of Fig. 7-5, obtained in a room having a reverberation time of 0.2 second, a broken line corresponding to RT60 = 0.18 second is tangent. Observations of this type have led some researchers to suspect that these subjective, psychoacoustical experiments are only a fancy way to measure reverberation time! And they seem to be about right over a limited region. Such echo disturbance curves may be divided into four general regions, *1* less than 1 ms where directional effects are determined, *2* from 1

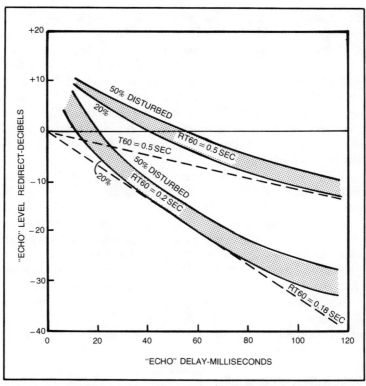

Fig. 7-5. The results of two psychoacoustical listening tests conducted by Nickson, Muncey, and Dubout (Ref. 49). Speech signals were presented to the subjects along with the same speech delayed varying amounts and at various levels. The tests were made in rooms having reverberation times of 0.5 and 0.2 second. The percentages of subjects disturbed by the echo are shown.

to about 50 ms where the integrating effect of the ear works, *3* an intermediate delay region largely dominated by reverberation, and *4* the long delay region where the curves approach the horizontal. For studios, control rooms, and listening rooms, we are primarily interested in *2* and *3*.

ECHOES IN SMALLER SPACES

Is it possible to hear echoes in an acoustically well treated studio, control room, or listening room? We do know that the room boundaries give rise to reflections which, in a physical sense at least, might be called echoes. We also know that sound recorded outdoors in the absence of such reflections has a flat, dead character. We like what we hear inside. The sound energy is contained by the boundaries of the room and the containment is directly responsible for not only increased loudness, but also the difference between indoor and outdoor sound quality.

Have you ever stopped to think as you watch a motion picture that you are sitting in pitch blackness a substantial fraction of the time? The eye-brain mechanism smooths out the intermittent pictures on the screen with an integrating action not unlike that of the ear-brain mechanism. In the Haas region the primary and delayed versions of the original sound are combined by our ear-brain mechanism.

This integrating action combines the energy of all the echo spikes arriving within about 50 ms, resulting in an increase in perceptual loudness and an improvement in quality. The "sound" of a studio, so sought after by contemporary musicians, is a direct result of this energy returned from room boundaries and integrated together by the ear-brain combination. We are conscious of the presence of these short delay "echoes" even though they are not discrete. They could be called incipient echoes (not "insipient" which means "stupid"!); echoes just coming into being. The "liveliness" of the room and the "body" and "substance" of the sound of a studio are the direct result of these short delay, incipient echoes.

REAL WORLD APPLICATION

We have sampled the results of sophisticated psychoacoustical tests, what do they have to do with, say, recording studio design and operation? Let's examine a specific application. Figure 7-6 shows echograms recorded in a 16,000 cu ft studio with acknowledged problems. Human ears had no trouble hearing the defect, but were less successful in pinpointing its source. Impulses recorded on tape were radiated from a loudspeaker into the studio. The original impulses were of nice, rectangular shape and had a duration of 1 ms. They were far from rectangular, however, after undergoing recording and playback on tape machines and the loudspeaker twisted the pulses out of shape some more, but at least the radiated clicks were crisp and short enough to delineate any echoes from surfaces differing more than a couple of feet. The microphone was placed successively at four random positions in the studio. The echo pattern characteristics of each of the four positions is shown in Fig. 7-6. In each case the arrival of the direct wave triggered the single sweep of the beam of the cathode ray oscilloscope. The vertical deflection is a linear arbitrary scale; the horizontal scale is 20 ms per division. Each spike represents a separate physical echo, although the ear integrates all spikes within about the first three horizontal scale markings (60 ms).

One thing immediately apparent from the echograms of Fig. 7-6 is that the echo pattern changes considerably from point to point in the studio. Our desire is now to evaluate the audible effects suggested by each echogram. One way to do this is to compare these echograms with the results of psychoacoustical tests made by others. By a stroke of good fortune (achieved by some sly maneuvering) we have just what is needed in Fig. 7-5. The reverberation time of the studio in which the echograms were recorded is 0.51 second (averaged 125-2000 Hz). This compares beautifully with the upper curves of Fig. 7-5. The 0.5 second curves of Fig. 7-5 have been reproduced in Fig. 7-7. The plotted dots, circles, plus signs, and triangular spots represent specific peaks from the echograms of Fig. 7-6. With the exception of microphone position M-3, practically all points fall below the 20% disturbed curve which means that they are probably inaudible. At position M-3, however, a number of peaks having delays greater than 50 ms fall well up in the disturbance region. The M-3 echogram is also unusual in that there is an echo at a delay of 20 ms that is 2.5 dB greater in amplitude than the direct pulse. This could come about only by (a) loudspeaker directivity, (b) an obstruction

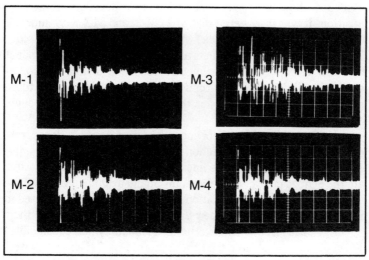

Fig. 7-6. Echograms taken at four different positions in a studio of 16,000 cu ft volume and having an average reverberation time of 0.51 second. The horizontal time scale is 20 ms/div.

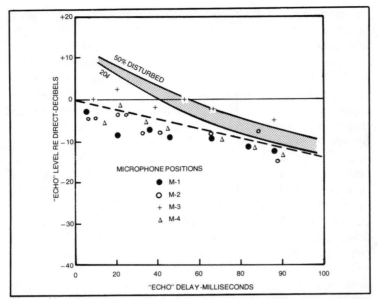

Fig. 7-7. The echo peaks of Fig. 7-6 are plotted with the 0.5 second curves of Fig. 7-5 to test for echo audibility. This test would indicate that echoes would be audible at microphone position M-3, but this is questioned in the text. The lone M-2 echo at 90 ms may also be audible, but less noticeable.

in the direct path, or (c) a focussing effect of some sort. With such evidence, the next step is to see what is so special about the M-3 position. It was discovered that M-1, M-2, and M-4 were more or less in front of the loudspeaker, while M-3 position was to one side. Loudspeaker directional effects undoubtedly discriminated against the M-3 direct pulse. The impulse response of this studio, as shown by Fig. 7-7 (disregarding the M-3 points), appears to be normal and the reasons for the complaints must be sought elsewhere.

In Fig. 7-7 the broken line again represents a decay rate corresponding to a reverberation time of 0.5 second. Note how the spots cluster just below this line. Well, why shouldn't they? After all, the echograms of Fig. 7-6 are only highly magnified views of the first 100 ms of the reverberation decay. The method illustrated in Fig. 7-7 emphasizes the idea that when echo peaks raise their "pointy" heads above the reverberation line they become audible. We could have evaluated the echograms of Fig. 7-6 almost as well by using the broken reverberation line of Fig. 7-7 rather than the shaded psychoacoustical curve, particularly above 50 or 60 ms. Not that we dislike psychoacousticians, but that we love simplicity!

130

Reverberation 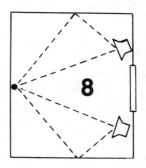 8

Is reverberation a good thing or bad? True friend or arch enemy? The dual effects of reverberation are akin to almost everything else in life: good in moderate quantities, bad in excess. In the preceding chapter we found that echoes may enhance sound indoors, but if amplitude and delay are excessive, they disturb. We also considered the fact that echoes are the very stuff of reverberation, that an echogram is but a detailed, expanded view of the early part of the reverberation decay. In this chapter we shall consider reverberation in its entirety, until a transient sound has died to the point of being engulfed in background noise or by another transient burst of speech or music following on its heels.

Reverberation may be considered good or bad depending on its degree or on the circumstances. Nowhere is the dichotomy of reverberation more clear than in its effect on music. In England a symphony orchestra was recorded as it played in a large anechoic chamber. The recording, made for research purposes, sounded horrible - even thinner, weaker, and less resonant than the usual outdoor sound of symphonic music because even reflections from the ground, trees, and stage backing were missing. Clearly, symphonic and other music requires reverberation, the amount depends on numerous factors to be considered in detail.

The effect of reverberation on the understandability of speech is similar to that on music in that too little as well as too much is harmful and in between there must be something approaching an optimum amount. The unamplified human voice outdoors cannot cover a very large audience and it sounds thin and lacks resonance, even as music. Too much reverberation, in the sense that it hangs on too long, confuses the consonants of speech upon which intelligibility depends. This is true of both amplified and un-amplified speech indoors.

GROWTH OF SOUND IN A ROOM

Referring to Fig. 8-1A, let us consider a source S and a human receiver H in a room. As source S is suddenly energized, sound travels outward from S in all directions. Sound travels a direct path to H and we shall consider it as time equals 0, as in Fig. 8-1B, that time at which the direct sound reaches the ears of listener H. The sound pressure at H instantly jumps to a value which is less than that which left S due to spherical divergence and small losses in the air. The sound pressure at H stays at this value until reflection R_1 arrives and then suddenly jumps to the $D+R_1$ value. Shortly thereafter R_2 arrives causing the sound pressure to increase a bit more. The arrival of each successive reflected component causes the level of sound to increase stepwise. These additions are, in reality, vector additions involving both magnitude and phase, but we are keeping things simple for this illustration.

Sound pressure at receiver H grows step by step as one reflected component after another adds to the direct component. The reason the sound pressure at H does not instantly go to its final value is that sound travels by paths of varying length. Although 1130 ft/sec is about the muzzle velocity of a .22 caliber rifle, reflected components are delayed an amount proportional to the difference in distance between the reflected path and the direct path. The buildup of sound in a room is thus relatively slow due to finite transit time.

The ultimate level of sound in the room is determined by the energy going into the source S. The energy it radiates (less than the input by the amount of loss in S) is dissipated as heat in wall reflections and other boundary losses and a small amount in the air itself. With a constant input to S, the sound pressure level builds up as in Fig. 8-1B to a steady state equilibrium, even as an automobile traveling steadily at 50 miles per hour with the accelerator in a given position. Pushing down on the accelerator pedal increases the energy to the engine and the automobile stabilizes at a new equilibrium point at which the many frictional losses are just supplied. Increasing the input to the source S means a new

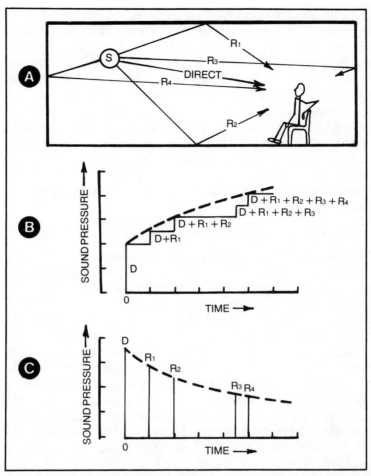

Fig. 8-1. The build-up and decay of sound in a room; (A) the direct sound arrives first at time = 0, the reflected components arriving later, (B) the sound pressure at H builds up stepwise, and (C) the sound decays exponentially after the source ceases.

equilibrium of room sound pressure level as room losses are just supplied.

DECAY OF SOUND IN A ROOM

Opening the switch feeding source S, the room is momentarily still filled with sound, but stability is destroyed because the losses are no longer supplied with energy from S. Rays of sound, however, are caught in the act of darting about the room with their support cut off.

What is the destiny of, for instance, the ceiling reflected component R_1? As S is cut off, R_1 is on its way to the ceiling. It loses energy at the ceiling bounce and heads toward H. Passing H it hits the rear wall, then the floor, the ceiling, the front wall, the floor again and so on . . . Losing energy at each reflection and spreading out all the time, it is soon so weak as to be considered dead. The same thing happens to R_2, R_3, R_4, and a multitude of others not shown. Figure 8-1C shows the exponential decrease of the first bounce components which would also apply to the wall reflections not shown and to the many multiple bounce components. The sound in the room thus dies away, but it takes a finite time to do so because of the speed of sound, losses at reflections, in the air, and in divergence.

IDEALIZED GROWTH AND DECAY OF SOUND

From the view of geometrical (ray) acoustics, decay of sound in a room as well as its growth is a stepwise phenomenon. However, in the practical world the great number of small steps involved result in smooth growth and decay of sound. In Fig. 8-2A the idealized forms of growth and decay of sound in a room are shown. Here the sound pressure is on a linear scale and is plotted against time. Figure 8-2B is the same thing except that the vertical scale is plotted in decibels, i.e., to a logarithmic scale.

During the growth of sound in the room power is being fed to the sound source. During decay the power to the source is cut off, hence the difference in the shapes of the growth and decay curves. The decay of Fig. 8-2B is a straight line in this idealized form and this becomes the basis for measuring the reverberation time of an enclosure.

REVERBERATION TIME

Reverberation time is defined as that time required for the sound in a room to decay 60 dB. That represents a change in sound intensity of 1 million or a change in sound pressure of 100,000. In very rough human terms it is approximately the time required for a sound that is very loud to decay to inaudibility. W.C. Sabine, the Harvard pioneer in acoustics who introduced the concept, used a portable windchest and organ pipes as a sound source, a stopwatch and a pair of keen ears to measure the time from the interruption of the source to inaudibility. Today we have better technical measuring facilities, but we can only refine our understanding of the basic concept Sabine gave us.

The approach to measuring reverberation time is illustrated in Fig. 8-3A. With some recording device which gives us a hardcopy trace of the decay it is a simple step to measuring the time required for the 60 dB decay. It is simple in theory, at least. In practice many problems are encountered. For example, obtaining a nice straight decay spanning 60 dB or more as in Fig. 8-3A is a very difficult practical problem. Background noise, an inescapable fact of life, suggests that a higher source level is needed. This may be a

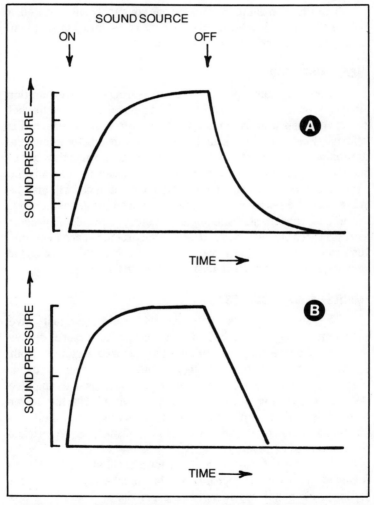

Fig. 8-2. The growth and decay of sound in a room; (A) vertical scale in linear sound pressure units, and (B) the vertical scale in logarithmic units, decibels.

possibility if the background noise level is 30 dB as in Fig. 8-3A as source levels of 100 dB are quite attainable. If, however, the noise level is near 60 dB as in Fig. 8-3B, a source level greater than 120 dB is required. If a 100 watt amplifier driving a certain loudspeaker gives a sound pressure level (SPL) of 100 dB at the required distance, doubling the power of the source increases the SPL only 3 dB, hence 200 watts gives 103 dB, 400 watts gives 106 dB, 800 watts gives 109 dB, etc. Limitations of size and cost can place a ceiling on maximum SPLs in a practical case.

The situation of Fig. 8-3B is the one commonly encountered, a usable trace less than the 60 dB desired. The solution is simply to extrapolate the straight portion of the decay.

MEASURING RT60

There are many approaches to measuring the reverberation time of a room and many instant readout devices are on the market to serve those who have only a casual interest in reverberation effects. For example, sound contractors need to know the approximate reverberation time of the spaces in which they are to install a sound reinforcement system and measuring it avoids the tedious process of calculating it. The measurements will also be more accurate because of uncertainty in absorption coefficients. Acoustical consultants called upon to correct a problem space or verify a carefully designed and newly constructed space generally lean toward the old fashioned method of recording many sound decays. These give much detail meaningful to the practiced eye.

IMPULSE SOUND SOURCES

The sound sources to excite the enclosure must have sufficient energy throughout the spectrum to assure decays sufficiently above the noise to give the required accuracy. Both impulse sources and those giving a steady state output are used. For large spaces, even small cannons have been used as impulse sources to provide adequate energy, especially in the lower frequencies. More common impulse sources are powerful electrical spark discharges and pistols firing blanks. Even pricked balloons have been used.

The impulse decays of Fig. 8-4 for a small studio have been included to show their appearance. The sound source used was a Japanese air pistol which ruptures paper discs. This pistol was originally intended as an athletic starter pistol but failed to find acceptance in that area. As reported by Sony engineers,[50] the peak

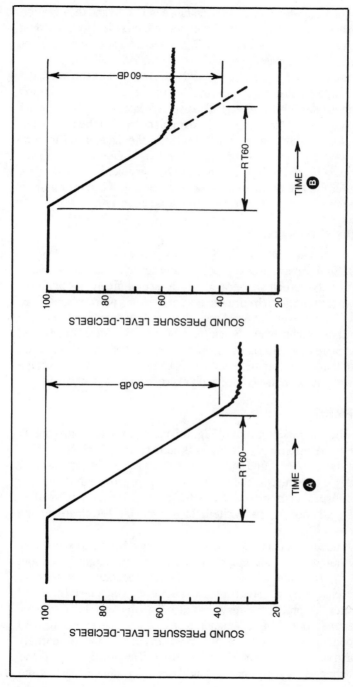

Fig. 8-3. The length of the decay is dependent on strength of source and noise level. (A) Rarely do practical circumstances allow a full 60 dB decay. (B) The slope of the limited decay is extrapolated to determine the reverberation time.

sound pressure level at 1 meter distance is 144 dB and the duration of the major pulse is less than 1 millisecond. It is ideal for recording echograms, in fact, the decays of Fig. 8-4 were made from impulses recorded for that very purpose.

In Fig. 8-4 the straight, upward traveling part on the left is the same slope for all decays because it is a result of machine limitation (writing speed 500 mm/sec.). The useful measure of reverberation is the downward traveling, more irregular slope on the right side. This slope yields reverberation time after the manner of Fig. 8-3. Note that the octave band noise level is higher for the lower frequency bands. In fact, the impulse barely poked its head above noise for the 250 Hz and lower octaves. This is a major limitation of the method unless the heavy artillery is rolled out.

STEADY STATE SOURCES

It has been noted that Sabine used a windchest and organ pipes. Sine wave sources, providing energy at a single frequency, give highly irregular decays which are difficult to analyze. Warbling a tone, which spreads its energy over a narrow band, is an improvement over the fixed tone, but random noise sources have essentially taken over. Bands of random noise give steady and dependable indication of the average acoustical effects taking place within that particular slice of the spectrum. Octave and 1/3 octave bands of random noise (white or pink) are most commonly used.

EQUIPMENT

The equipment layout of Fig. 8-5 used by the author to obtain the reverberation decays to follow is quite typical. A wideband pink noise signal is amplified and used to drive a rugged loudspeaker. A switch for interrupting the noise excitation is provided. By aiming the loudspeaker into a corner of the room (especially in smaller rooms) all modes are excited, for all modes terminate in the corners.

A nondirectional microphone is positioned on a tripod, usually at ear height for a listening room or microphone height for a room used for recording. The smaller the microphone, the less its directional effects. Some of the larger microphones (e.g., 1″ diameter diaphragms) may be fitted with random incidence correctors, but using a smaller (e.g., ½″ diameter diaphragm) microphone is considered the best insurance for essentially uniform sensitivity to sound arriving from all angles. In Fig. 8-5 the microphone is a high quality condenser microphone, part of the

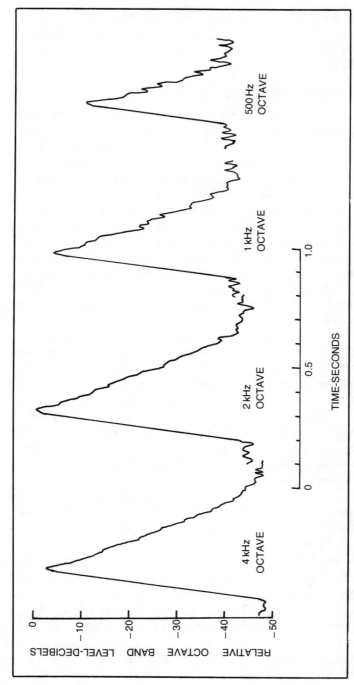

Fig. 8-4. Reverberatory decays resulting from impulse excitation of a small studio. The upgoing left side of each trace is recording machine limited, the downgoing right side is the reverberatory decay.

139

Brüel & Kjaer 2215 sound level meter, but separated from it by an extension cable. This provides an excellent preamplifier, built in octave filters, a calibrated system, and a line level output signal for the tape recorder.

MEASUREMENT PROCEDURE

Every time the switch is closed the room is filled with a very loud wideband pink noise "sh-h-h-h" sound. This is usually loud enough to require using earmuffs for all in the room. Opening the switch, the sound in the room decays. The microphone, at its selected position, picks up this decay which is recorded on magnetic tape for later analysis and study.

Signal to noise ratio determines the lengths of the reverberatory decay available for study. As mentioned, it is rarely possible to realize the entire 60 dB decay involved in the definition of RT60, nor is it necessary. It is quite possible, however, to get 45 to 50 dB decays with the equipment shown in Fig. 8-5 by the simple expedient of double filtering. For example, the octave filter centered on 500 Hz in the sound level meter is used both in recording and in later playback for analysis.

The analysis procedure outlined in the lower part of Fig. 8-5 utilizes the same magnetic recorder and B&K 2215 sound level meter with the addition of a B&K 2305 graphic level recorder. The line output of the tape recorder is connected to the front end of the sound level meter circuit through a 40 dB attenuating pad. To do this the microphone of the sound level meter is removed and a special fitting screwed in its place. The output of the sound level meter is plugged directly into the graphic level recorder input, completing the equipment interconnection. The appropriate octave filters are switched in as the played back decay is recorded on the level recorder. The paper drive provides the spreading out of the time dimension at adjustable rates. The graphic level recorder offers a 50 dB recording range for the tracing pen on the paper.

ANALYSIS OF DECAY TRACES

An octave slice of pink noise viewed on a cathode ray oscilloscope shows a trace that looks very much like a sine wave except that it is constantly shifting in amplitude and phase. Well, that was the definition of random noise in Chapter 1. This characteristic of random noise has its effect on the shape of the reverberatory decay trace. Consider what this constantly shifting random noise signal does to the normal modes of a room. When

140

axial, tangential, and oblique modes are considered, they are quite close together on the frequency scale. The number of modes included within an octave band centered on 63 Hz in the specific case to be elaborated later is as follows: 4 axial, 6 tangential, and 2 oblique modes between the −3 dB points. These are graphically shown in Fig. 8-6 in which the taller lines represent the more potent axial modes, the intermediate height the tangential modes, and the shorter lines the oblique modes.

As the switch of Fig. 8-5 is closed, the high level random noise from the loudspeaker energizes the various modes of the room, hitting mode A and an instant later hitting mode B. While the shift is being made in the direction of mode B, mode A begins to decay. Before it decays very far, however, the random noise instantaneous frequency is once more back on A, giving it another boost. All the modes of the room are in constant agitation, alternating between high and somewhat lower levels, as they start to decay in between kicks from the loudspeaker.

At what point will this erratic dance of the modes be as the switch is opened to begin the decay? It is strictly a random situation, but it can be said with confidence that each time the switch is opened for 5 successive decays, the modal excitation pattern will be somewhat different each time. The 12 modes in the 63 Hz octave will all be highly energized, but each to a somewhat different level the instant the switch is opened.

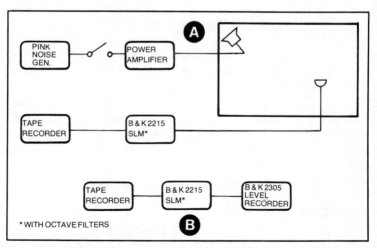

Fig. 8-5. Equipment arrangement for measuring reverberation time of an enclosure; (A) recording decays on tape on location, (B) later recording decays for analysis.

MODE DECAY VARIATIONS

To endow this discussion with a glow of veracity, real life measurements in a real room will be discussed. The room is a rectangular studio for voice recording having the dimensions 20′ 6″ x 15′0″ x 9′6″, with a volume of 2,921 cu.ft. The measuring equipment is exactly that outlined in Fig. 8-5 and the technique is that described above. Four successive 63 Hz octave decays traced directly from the graphic level recorder paper are shown in Fig. 8-7A. These traces are not identical and any differences must be attributed to the random nature of the noise signal because everything else was held constant. The fluctuations in the decays result from beats between closely spaced modes. Because the excitation level of the modes is constantly shifting, the form and degree of the beat pattern shifts from one decay to another depending on where the random excitation happens to be the instant the switch is opened. Even though there is a family resemblance between the four decays, fitting a straight line to evaluate the RT60 of each can well be affected by the beat pattern. For this reason I make a practice of recording five decays for each octave for each microphone position of a room. With 8 octaves (63 Hz-8 kHz), 5 decays per octave, and 3 microphone positions, this means 120 separate decays to fit and figure for each room, which is laborious. This approach is one way to get a good, statistically significant view of the variation with frequency. A different system could accomplish this with less work, but most of them do not give hard copy detail of the shape of each decay. There is much information in each decay and acoustical flaws can often be identified from aberrant decay shapes.

Four decays at 500 Hz are also shown in Fig. 8-7 for the same room and the same microphone position. The 500 Hz octave (354 - 707 Hz) embraces about 2,500 room modes. With such a high mode density the 500 Hz octave decay is much smoother than the 63 Hz octave with only a dozen. Even so, the irregularities for the 500 Hz decay of Fig. 8-7 are due to the same cause. Remembering that some modes die away faster than others, the decays in Fig. 8-7 for both octaves are a composite of all modal decays involved.

WRITING SPEED

The B&K 2305 graphic level recorder is capable of wide adjustment of writing speed. A sluggish pen response is useful when fast fluctuations need to be ironed out. When detail is desired, faster writing speeds are required. Inasmuch as too slow

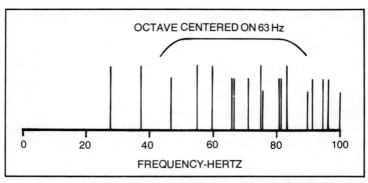

Fig. 8-6. The normal modes included (−3 dB points) in an octave centered on 63 Hz. Tallest lines represent axial modes, intermediate length tangential modes, and shortest ones oblique modes.

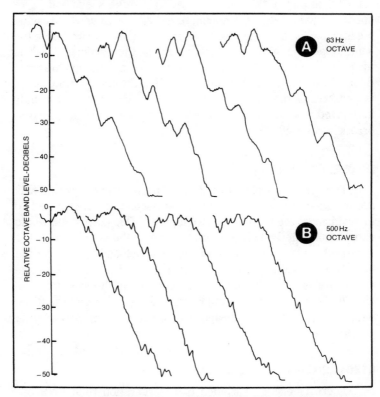

Fig. 8-7. Actual decays of random noise recorded in a small studio having a volume of 2,921 cu. ft.; (A) four successive 63 Hz octave decays recorded under identical conditions, (B) four successive 500 Hz octave decays also recorded under identical conditions. The differences noted are the differences in random noise excitation the instant the switch is opened beginning the decay.

writing speed might affect the rate of decay as it smooths out the trace, we must look into it a bit.

In Fig. 8-8 the same 63 Hz decay is recorded with five different pen response speeds ranging from 200 to 1,000 mm/sec. The instrument limited decay for each is indicated by the solid straight lines. A writing speed of 200 mm/sec. really smooths the fluctuations. The decay detail increases as the writing speed is increased, suggesting that a cathode ray oscilloscope tracing of the decay might show even more of modal interference effects during the decay.

The big question is, does writing speed affect the decay slope from which we read RT60 values? Obviously, an extremely slow pen response would record the machine's decay characteristic rather than that of the room. For every writing speed and paper speed setting there is a minimum reverberation time which can be measured. The broken lines drawn through the decays all have the same slope. In Fig. 8-8 it would appear that this particular RT60 is measured equally well by any of the five traces, although the more detail the more uncertainty in fitting a straight line. Writing speed is just one of the several adjustments which must be carefully monitored to assure that important information is not obscured or errors introduced.

FREQUENCY EFFECT

Typical decays of octave bands of noise from 63 Hz to 8 kHz are included in Fig. 8-9. The greatest fluctuations are in the two lowest bands, the least in the two highest. This is what we would expect from the knowledge that the higher the octave band, the greater the number of normal modes included, and the greater the statistical smoothing. We should not necessarily expect the same decay rate as reverberation time is different for different frequencies. In the particular voice studio case of Fig. 8-9, a uniform RT60 with frequency was the design goal which was approximated in practice.

RT60 VARIATION WITH POSITION

There is enough variation of reverberation time from one position to another in most rooms to justify taking measurements at several positions. The average then gives a better statistical picture of the behavior of the sound field in the room. If the room is symmetrical, it may be wise to spot all measuring points on one

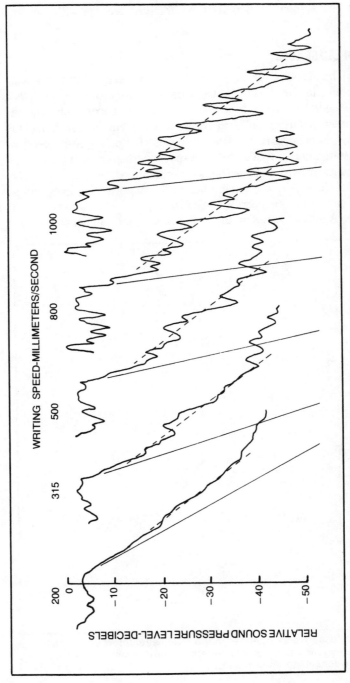

Fig. 8-8. The same 63 Hz octave decays recorded with five different recorder pen response speeds. The solid straight lines indicate instrument limited decay, the broken lines are all of the same slope.

145

side of the room to increase the effective coverage with a given effort.

COUPLED SPACES

The shape of the reverberation decay can point to acoustical problems in the space. One common effect which alters the shape of the decay is that due to acoustically coupled spaces. This is quite common in large public gathering spaces, but is also found in offices, homes, and other smaller spaces. The principle involved is illustrated in Fig. 8-10. The main space, perhaps an auditorium, is acoustically quite dead and has a reverberation time corresponding to the slope A. An adjoining hall with very hard surfaces and a reverberation time corresponding to slope B opens into the main room. A person seated in the main hall near the hall opening could

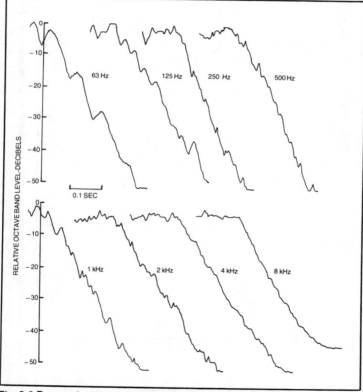

Fig. 8-9. Decay of octave bands of noise in a small studio of volume 2,921 cu. ft.. Fluctuations due to modal interference are greatest for low frequency octaves containing fewer modes.

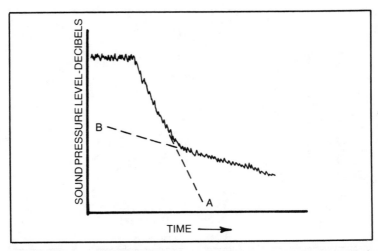

Fig. 8-10. Reverberatory decay of double slope due to acoustically coupled spaces. The shorter reverberation time represented by slope A is that of the main room. A second, highly reflective space is coupled through an open doorway. Those seated near the doorway are subjected first to main room response and then the decay of the coupled space.

very well experience a double slope reverberation decay. Not until the sound level in the main room falls to a fairly low level would the main room reverberation be dominated by sound fed into it from the slowly decaying sound in the hall. Assuming slope A is correct for the main room, persons subjected to slope B would hear greatly inferior sound.

INFLUENCE OF REVERBERATION ON SPEECH

Let us think about what happens to just one tiny word in a reverberant space. The word is "back". It starts abruptly with a "ba . . . " sound and ends with the consonant " . . . ck" which is much lower in level. As measured on the graphic level recorder, the "ck" sound is about 25 dB below the peak level of the "ba" sound and reaches a peak 320 milliseconds after the "ba" peak.

Both the "ba" and the "ck" sounds are transients which build up and decay after the manner of Fig. 8-2. Sketching these various factors to scale yields something like Fig. 8-11. The "ba" sound builds to a peak at an arbitrary level of 0 dB at time=0, after which it decays according to the reverberation time of the room which is assumed to be 0.5 second (60 dB decay in 0.5 second). The "ck" consonant sound, occuring 0.32 second later, peaks 25 dB below the "ba" sound peak. It too decays at the same rate as the "ba"

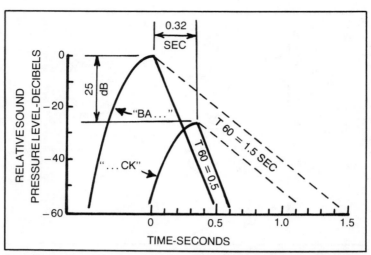

Fig. 8-11. An illustration of the effects of reverberation on intelligibility of speech. Understanding the word "back" depends on apprehending the later, lower level consonant " . . . ck" which is masked by reverberation if the reverberation time is too long.

sound according to the 0.5 second law assumed. Under the influence of the 0.5 second reverberation time, the "ck" consonant sound is not masked by the reverberation time of "ba". If the reverberation time is increased to 1.5 second, as shown by the broken lines, the consonant "ck" is covered completely by reverberation.

The primary effect of excessive reverberation is to impair the intelligibility of speech by masking the lower level consonants. In the word "back", without a clear grasping of the "ck" part, the meaning of the word is lost. Understanding the "ck" ending is the only way to distinguish "back" from bat, bad, bass, ban, or bath. In this oversimplified way we can grasp the effect of reverberation on the understandability of speech and the reason why speech is more intelligible in rooms having lower reverberation times.

Sound reinforcement engineers have been assisted greatly by the work of the Dutch investigators, Puetz[51] and Klein[52]. Through their work it is a straightforward procedure to predict with reasonable accuracy the intelligibility of speech in a space from geometrical factors and a knowledge of reverberation time.

INFLUENCE OF REVERBERATION ON MUSIC

The effect of hall "resonance" or reverberation on music is intuitively grasped, but not generally deeply understood in its more complicated aspects. This is a subject which has received much attention from scientists as well as musicians and a final word has not yet appeared. Beranek has made a valiant attempt to summarize present knowledge and to pinpoint essential features of music halls around the world,[53] but our understanding of the problem is still quite incomplete. Suffice it to say that the reverberation decay of a music hall is only one important factor among many, another being the echo pattern, especially the so-called "early sound". It is beyond the scope of this book to treat this subject in any detail, but an interesting point or two, commonly overlooked, will be mentioned briefly.

We have considered normal modes in some detail because of their basic importance. They are also, of course, active in music halls and listening rooms. An interesting phenomenon is pitch change during reverberant decay. In reverberant churches organ tones have been observed to change pitch as much as a semitone during decay. In searching for an explanation, two things have been mentioned; (1) shift of energy between normal modes, and (2) the perceptual dependence of pitch on sound intensity. There are problems in both. Balachandran has demonstrated the physical (as opposed to pyschophysical) reality of the effect[54] by analyzing by Fast Fourier Transform (FFT) technique the reverberant field created by 2 kHz pulses. He revealed the existence of a primary

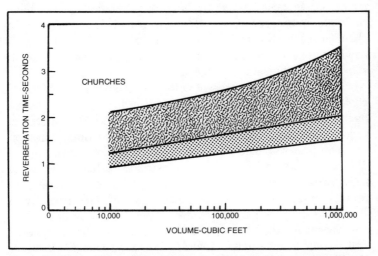

Fig. 8-12. "Optimum" reverberation time for churches. Upper area applies to more reverberant liturgical churches and cathedrals, lower area to churches having more speech oriented services. A compromise between music and speech is required for most churches.

149

1,992 Hz spectral peak and, curiously, another peak at 3,945 Hz. As a 6 Hz change would be just perceptible at 2 kHz and a 12 Hz change at 4 kHz, we see that the 39 Hz shift from the octave of 1,992 Hz would give a definite impression of pitch change. The reasons for this are still under study. The reverberation time of the hall in which this effect was recorded is about 2 seconds.

OPTIMUM REVERBERATION TIME

From an elementary viewpoint, there must be some form of optimum reverberation time between the "too flat" condition of the outdoors and anechoic chambers and the obvious problems associated with excessively long reverberation times of a stone cathedral. Well, such an optimum does exist, but there is usually great disagreement as to just what it is. This is a subjective problem and we must expect some differences in opinions. The optimum value depends on the sounds being considered.

Reverberation rooms, used for measuring absorption coefficients, are carefully designed for the longest practical RT60 as this is related to the accuracy of the measurements made in them. The optimum here is the maximum attainable.

The best reverberation time for a space in which music is played for the benefit of listeners depends on the size of the space and the type of music. Slow, solemn, melodic music, such as that of an organ, is best served by long reverberation time. Quick rhythmic music requires a different reverberation time than chamber music. No single optimum, obviously, will fit all types of music, the best that can be done is to establish a range based on subjective judgements of specialists.

Recording studios present still other problems which do not conform to simple rules. Separation recording in which instruments are recorded on separate tracks for later mixdown in general require quite dead spaces to realize adequate acoustical separation between tracks. Music directors and band leaders often require different reverberation for different instruments, hence hard areas and absorptive areas may be found in the same studio. The range of reverberation realized in this manner is limited, but proximity to absorptive or reflective surfaces does affect local conditions.

Spaces for speech, as we have seen, require shorter reverberation time than for music as we are interested primarily in direct sound. In general long reverberation time tends toward lack of definition and clarity in music and loss of intelligibility in speech. In dead spaces in which reverberation time is very short, loudness

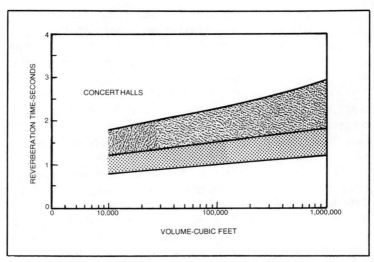

Fig. 8-13. "Optimum" reverberation time for concert halls; symphony orchestras near top of shaded area, lighter music lower. The lower shaded area applies to opera and chamber music.

and tonal balance may suffer. It is not possible to be specific in specification of optimum reverberation times for different services, but Figs. 8-12 through 8-14 are at least a rough indication of recommendations given by a host of experts in the field who do not always agree with each other.

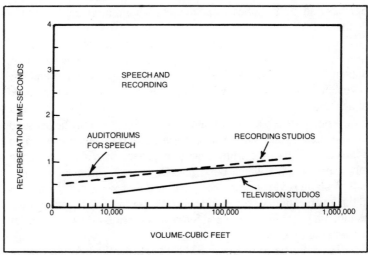

Fig. 8-14. Spaces designed for speech and recording require shorter reverberation times.

The reverberation times for churches in Fig. 8-12 range from highly reverberant liturgical churches and cathedrals to the shorter ranges of the lower shaded area characteristic of the more talk oriented churches. Churches generally represent a compromise between music and speech.

Figure 8-13 represents the range of recommended reverberation times for concert halls of various kinds. Those for symphony orchestras are near the top, those for lighter music somewhat lower. The lower shaded area applies for opera and chamber music.

Those spaces used primarily for speech and recording require close to the same reverberation times as shown in Fig. 8-14. Television studios are of even shorter reverberation time to deaden production sounds associated with rolling cameras, dragging cables, etc. It should also be remembered that local acoustics in television are dominated by the settings and local furnishings. In many of the spaces represented in Fig. 8-14 speech reinforcement is employed.

ARTIFICIAL REVERBERATION

Sometimes it is desirable to add artificial reverberation (not "echo") to the program signal when the natural reverberation of the recording space is insufficient. There are many ways of creating such reverberation. Unfortunately, the simpler ones are plagued with colorations and other quality and operational problems. Springs and multiple head tape devices have limited application in quality work.

The first, and perhaps one of the most satisfactory reverberation arrangements is the reverberation room. The program is played into this room and is picked up by a microphone and the reverberated signal is mixed into the original in an amount required to achieve the desired result. Such rooms may be especially built for the purpose[55,56] or, an available room or hallway may be pressed into service if sufficiently reflective and if extraneous noise can be controlled. This three dimensional approach provides the dense mixture of modes which best imitates reverberation of a larger hall. A small reverberation room has widely spaced normal modes in the low frequency region and colorations resulting therefrom are as much a problem when used for reverberation as when used for listening or recording.

The reverberation plate (EMT-140TS, Gotham Audio Corporation) dominates the field because of its modest size and cost (compared to a room), clean bass, and controllable reverberation time. Bending vibrations imparted to the plate are picked up at another point by a piezoelectric accelerometer attached to it. The decay is adjusted by bringing a sheet of porous material close to the metal sheet, adding dissipation. The older, large plate, enclosed in a box 8 x 4 x 1 ft has served and is still serving well, but a newer plate (EMT-240) device utilizing a special alloy of gold foil is replacing it. The application of digital techniques is initiating a whole new flood of devices giving greater control and excellent operating characteristics.

SABINE EQUATION

Sabine's reverberation equation was developed at the turn of the century in a strictly empirical fashion. He had several rooms at his disposal and by adding or removing seat cushions of a uniform kind he established the following relationship (adapted from the metric units he used):

$$RT60 = \frac{0.049\,V}{S\,a} \qquad (8\text{-}1)$$

where

$$
\begin{aligned}
TR60 &= \text{reverberation time, seconds} \\
V &= \text{volume of room, cu ft} \\
S &= \text{total surface area of room, sq ft} \\
a &= \text{average absorption coefficient of} \\
&\quad \text{room surfaces, sabin per sq ft} \\
S\,a &= \text{total Sabine absorption, sabins.}
\end{aligned}
$$

EYRING EQUATION

Another reverberation equation, an extension of the Sabine equation and extensively advocated, is the Eyring equation:[57,58]

$$RT60 = \frac{0.049\,V}{-S\log_e(1-\alpha)} \qquad (8\text{-}2)$$

where α is an average sound energy absorption coefficient of the surface. All of the factors of this equation are the same as for Equation 8-1 except the absorption coefficient. Young[59] has pointed out that the absorption coefficients published by materials man-ufacturers (such as the list in the appendix) are Sabine coefficients and can only be directly applied in the Sabine equation. Young recommends, after a thorough study of the historical development of the offshoots of Sabine's work, that Equation 8-1 should be used for all engineering computations. Two unassailable reasons for this are simplicity and consistency. In spite of the fact that this simpler procedure was suggested as early as 1932, and Young's convincing arguments for it were given in 1959, many technical writings (including the author's) have continued to put forth the Eyring equation for studio use. Even though there was authoritative backing for boosting Eyring for more absorbent spaces, why continue if the commonly available coefficients apply only to Sabine? These are the reasons for use of only Equation 8-1 in the illustrative reverberation calculations to follow.

The total Sabine absorption in a room would be easy to get if all surfaces of the room were uniformly absorbing but this condition rarely exists. Walls, floor, and ceiling may well be covered with quite different materials and then there are the doors and windows. The total absorption, Sa, of Equation 8-1 may be found by considering the absorption contributed by each surface. For example, in our imaginary room, let us say that an area S_1 is covered with a material having an absorption coefficient a_1 as obtained from the table in the appendix. This area then contributes $(s_1)(a_1)$ absorption units, called sabins, to the room. Likewise, another area S_2 is covered with another kind of material with absorption coefficient a_2 and it contributes $(S_2)(a_2)$ sabins of absorption to the room. The total absorption in the room is $S a = S_1a_1 + S_2a_2 + S_3a_3 \ldots$ etc. With a figure for Sa in hand, it is a simple matter to go back to Equation 8-1 and calculate the reverberation time, RT60.

REVERBERATION CALCULATION

To illustrate the process of calculating the reverberation time of an enclosure, let us apply what we have just discussed to a room 15 x 20 x 8 ft in size. Its volume is then $(15)(20)(8) = 2400$ cu ft. Its surface area is the sum of the areas of the 2 ends, 2 side walls, and ceiling and floor or $(2)(8)(15) + (2)(8)(20) + (2)(15)(20) = 1160$ sq ft.

Absorption coefficients for different materials are to be found in the appendix. For our example we assume that the floor is covered with heavy carpet, the walls and ceiling are of ½″ gypsum board, and there is no further treatment in the room. There is one 2′ 6″ x 6′ 8″ door and a 3′ x 5′ glass window. To simplify, the calculations will be carried out only for 500 Hz, although reverberation time for other frequencies is usually required. The next step is to build a table:

Material	S Area, sq ft	a absorption coefficient	S a Absorption units, sabins
Carpet, heavy	300	0.30	90.0
Door, wooden	17	0.10	1.7
Window	15	0.04	0.6
Gypsum board, ½″	828	0.05	41.4
	1160		133.7

Substituting known values in Equation 8-1:

$$RT60 = \frac{0.049\,V}{S\,a}$$

$$= \frac{(0.049)\,(2400)}{(133.7)}$$

$$= 0.88 \text{ second}$$

For examples of calculations involved in actual studio design the reader is referred to another of my books which includes details on a dozen specific studio designs.[109]

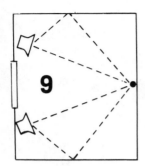

9

Common Signals — Noise, Speech, and Music

And in that order. The simple first, building to the complex. The word signal implies that some information is being conveyed. How can noise be considered an information carrier? An enthusiastic "Bronx cheer" razzberry conveys considerable information on the perpetrator's attitude toward someone or something. Noise is the basic part of such a communication, modulated in just the right way. Interrupting noise to form dots and dashes would be a way to shape noise into communication. We have also seen how a band of noise decaying can give information on the acoustical quality of a room. Of course, there is that which we call noise which is undesirable sound. Sometimes it is difficult to tell whether it is the unpleasant thing we call noise or a carrier of information. I don't particularly care for the noise my automobile makes, but it does convey considerable information on how well it is running. One person's noise may be someone else's communication. A hi-fi system may produce some beautiful sounds deemed very desirable by the owner, but to a neighbor, at 2 AM, they may not be considered beautiful at all. Sometimes it may not be easy to distinguish between information and noise. The same sound may be both. Society establishes limits to keep objectionable noise to a minimum while insuring that information carrying sounds can be heard by those who need to hear them.

ARTIFICIAL LARYNX

Noise which contains energy over a wide range of constantly shifting frequencies, phases, and amplitudes can even be shaped into speech. Sometimes people lose their voices for one reason or another. Perhaps the vocal cords are paralyzed or the larynx has been removed surgically. For these the Western Electric Company offers a prosthetic device which, held against the throat, produces pulses of sound which simulate the sounds produced by the natural vocal cords as they interrupt the air stream. This battery operated device even has a pitch control for controlling "voice" pitch. The tongue, lips, teeth, nasal passages, and throat then perform their normal function of molding the pulsed noise into words. Even if the overall effect leans somewhat in the Donald Duck direction, someone can speak who formerly could not and the "noise" has been shaped to some good end.

THE VOICE SYSTEM

One of the many amazing things about the human body is the high degree of efficiency associated with multiple use of organic systems. The functions of eating, breathing, and speaking all take place in relative simultaneous harmony. We can eat, breathe, and talk practically at the same time through the interworking of muscle action and valves without food going down the wrong hatch. If we sometimes try to do too many things at once the system is momentarily thwarted and we agonize as a bit of food is retrieved from the wrong pipe.

SOUND SPECTROGRAPH

An understanding of the sounds of speech is necessary to an understanding of how the sounds are produced. Speech is highly variable and transient in nature, comprising energy chasing up and down the three dimensional scales of frequency, intensity, and time. It takes the sound spectrograph to show all three on the same flat surface such as the pages of this book. Examples of several commonly experienced sounds revealed by the spectrograph are shown in Fig. 9-1. In these time progresses horizontally to the right, frequency increases from the origin upward, and the intensity is roughly indicated by the density of the trace - the blacker the trace the more intense the sound at that frequency and at that moment of time. Random noise on such a plot would show up as a gray, slightly mottle rectangle as all frequencies in the audible range and all intensities are represented as time progresses. The snare drum sound approaches random noise in spots, but is intermittent. The "wolf whistle", familiar to all red blooded male and female American types, opens on a rising note, a gap, and then a similar rising note which then falls in frequency as time goes on. The police whistle is a tone, slightly frequency modulated. Each

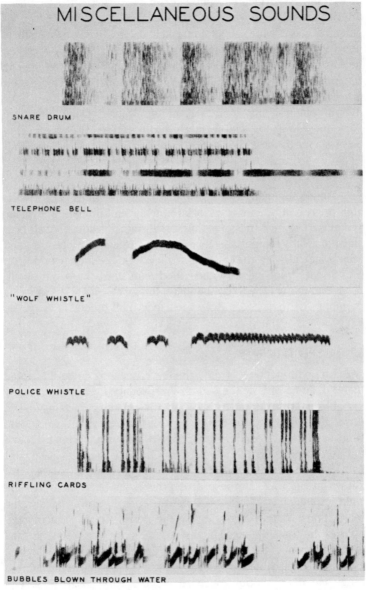

Fig. 9-1. Sound spectrographic recordings of common sounds. Time progresses to the right, the vertical scale is frequency, and intensity of components is represented by the intensity of the trace (courtesy of Bell Telephone Laboratories, Inc.).

VOICE SOUNDS

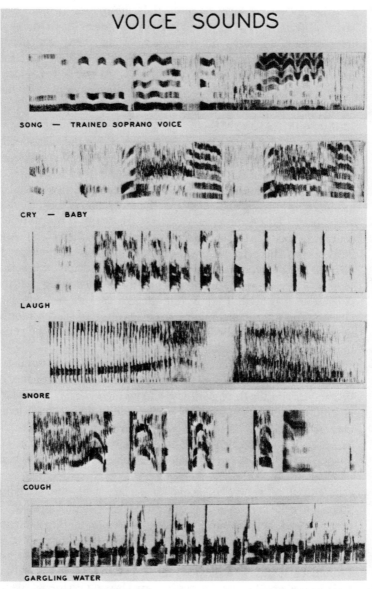

SONG — TRAINED SOPRANO VOICE

CRY — BABY

LAUGH

SNORE

COUGH

GARGLING WATER

Fig. 9-2. Sound spectrograms of human sounds other than speech (courtesy of Bell Telephone Laboratories, Inc.).

common noise has its spectrographic signature which reveals the very stuff which characterizes it.

The human voice mechanism is capable of producing many sounds other than speech. Figure 9-2 shows a number of these as revealed by sound spectrograms. It is interesting to note that harmonic trains appear on a spectrogram as more or less horizontal lines spaced vertically in frequency. These are particularly noticeable in the trained soprano's voice and the baby's cry (the association of the two is strictly fortuitous), but traces are evident in other spectrograms. The discussion to follow owes much to the clear presentation of Flanagan.[60]

SOUND SOURCES FOR SPEECH

The artificial larynx is based on the fact that there are really two more or less independent functions in the generation of speech sounds, the sound source and the vocal system. It is, in general, a series flow as pictured in Fig. 9-3A in which the raw sound is produced by a source and, after that, the raw sound is shaped in the vocal tract. To be more exact, there are really three different sources of sound to be shaped by the vocal tract as indicated in Fig. 9-3B. First, there is the one we naturally think of, the sounds emitted by the vocal cords. These are formed into the *voiced sounds*. They are produced by air from the lungs flowing past the slit between the vocal cords (the glottis) causing the cords to vibrate. The air stream, broken into pulses of air, produces a sound which can almost be called periodic, that is, repetitive in the sense that one sine wave cycle follows another.

The second source of sound is that made by forming a constriction at some point in the vocal tract with teeth, tongue, or lips and forcing air through it with a force high enough to produce significant turbulence. Air tumbling around in violent turbulence creates noise. This noise is shaped by the vocal tract to form the *fricative sounds* of speech such as the consonants f, s, v, and z. Try making these sounds and you will see that high velocity air is very much involved.

The third source of sound is produced by the complete stoppage of the breath, usually toward the front, a building up of the pressure and then the sudden release of the breath. Try speaking the consonants k, p, and t and you will sense the force of such plosive sounds. They are usually followed by a burst of fricative or turbulence sound. These three types of sounds, voiced, fricative, and plosive, are the raw sources that are shaped into the words we casually speak without giving a thought to the wonder of their formation.

VOCAL TRACT MOLDING OF SPEECH

The vocal tract can be considered as an acoustically resonant system. This tract, from the lips to the vocal cords, is about 6.7

inches (17 cm) long. Its cross sectional area is determined by placement of lips, jaw, tongue, and velum (a sort of trapdoor which can open or close off the nasal cavity) and varies from zero to about 3 sq. in. (20 sq. cm.). The nasal cavity is about 4.7 inches long (12 cm) and has a volume of about 3.7 cu. in. (60 cu. cm). These dimensions are mentioned because they have a bearing on the resonances of the vocal tract and their effect on speech sounds.

FORMATION OF VOICED SOUNDS

If the symbolic boxes of Fig. 9-3 are elaborated into source spectra and modulating functions, we arrive at something everyone in audio is interested in, the spectral distribution of energy in voice spectra. We also get a better understanding of the aspects of voice sounds which contribute to the intelligibility of speech in reverberation, noise, etc. Figure 9-4 shows the steps in producing voiced sounds. First, there is the sound produced by the vibration of the vocal cords, pulses of sound having a line spectrum which falls off at about 10 dB per octave as frequency is increased as shown in Fig. 9-4A. The sounds of the vocal cords pass through the vocal tract which acts as a filter varying with time. The humps of Fig. 9-4B are due to acoustical resonances (called formants) of the vocal pipe,

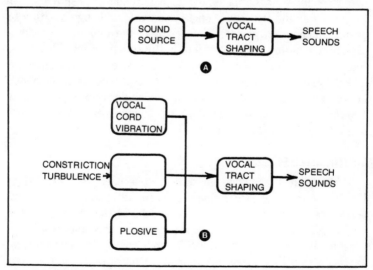

Fig. 9-3. (A) The human voice is produced through the interaction of two essentially independent functions, a sound source and a time varying filter action of the vocal tract. (B) The sound source can be broken down into vocal cord vibration for voiced sounds, the fricative sounds resulting from air turbulence, and the plosive sounds.

161

essentially open at the mouth end and essentially closed at the vocal cord end. Such an acoustical pipe 6.7 inches long has resonances at odd quarter wavelengths and these peaks occur at approximately 500, 1,500, and 2,500 Hz. The output sound, shaped by the resonances of the vocal tract, is shown in Fig. 9-4C. This applies to the voiced sounds of speech.

FORMATION OF UNVOICED SOUNDS

Unvoiced sounds are shaped in a similar manner as indicated in Fig. 9-5. In the unvoiced case we start with the distributed, almost random-noise-like, spectrum of the turbulent air as fricative sounds are produced. The distributed spectrum of Fig. 9-5A is generated near the mouth end of the vocal tract, rather than the vocal cord end, hence the resonances of Fig. 9-5B are of somewhat different shape. Figure 9-5C shows the sound output shaped by the time varying filter action of Fig. 9-5B.

PUTTING IT ALL TOGETHER

The voiced sounds, having their origin in vocal cord vibrations, and the unvoiced sounds originating in turbulences, and plosives, originating near the lips, go together to form all our speech sounds. As we speak, the formant resonances shift about in frequency as the lips, jaw, tongue, and velum change position to shape the desired words. The result is the unbelievable complexities of human speech evident in the spectrogram of Fig. 9-6. Information communicated via speech is a pattern of frequency and intensity shifting rapidly with time. Note that there is little speech energy above 4 kHz in Fig. 9-6 nor, which does not show, below 100 Hz. We understand now why the presence filter peaks in the 2 to 3 kHz region; that is where the pipes resonate!

SYNTHESIZED SPEECH

Mechanical speaking machines date back to 1779 when Kratzenstein of St. Petersburg constructed a set of acoustical resonators to emulate the human mouth. These were activated with reeds such as those of a mouth organ. He was able to produce reasonably recognizable vowel sounds with the contraption. Wolfgang von Kempelen of Vienna did a much better job in 1791 which Wheatstone later improved. This machine used a bellows to supply air to a leather tube which was manipulated by hand to simulate mouth action and included an "S" whistle, a "SH" whistle, and a nostril cutoff valve. After experimenting with a copy of

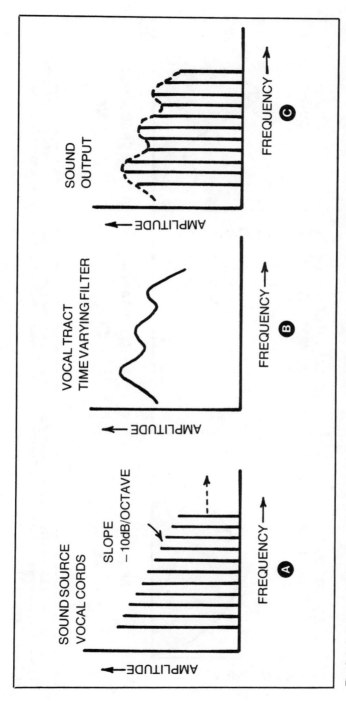

Fig. 9-4. A diagram of the production of voiced sounds such as vowels. (A) The vocal cord sound is primarily a line spectrum with a − 10 dB/octave slope. (B) The time varying filter action of the vocal tract shows resonances at odd quarter wavelengths of the tract acting as an acoustical pipe. (C) The filter action of (B) shapes the vocal cord sound of (A) to give the output sound.

163

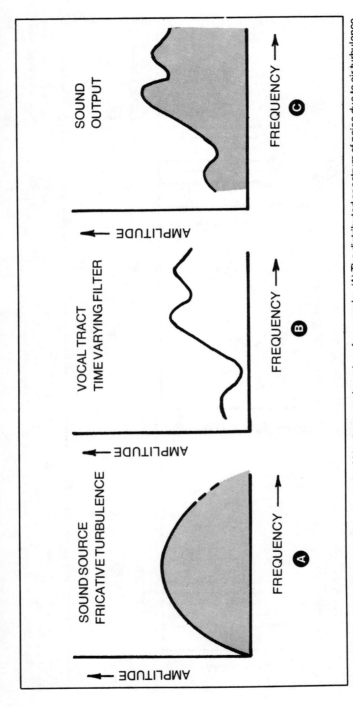

Fig. 9-5. A diagram of the production of unvoiced fricative sounds such as f, s, v, and z. (A) The distributed spectrum of noise due to air turbulence resulting from constrictions of the vocal tract. (B) The time varying filter action of the vocal tract. (C) The output sound resulting from the filter action on the distributed sound of (A).

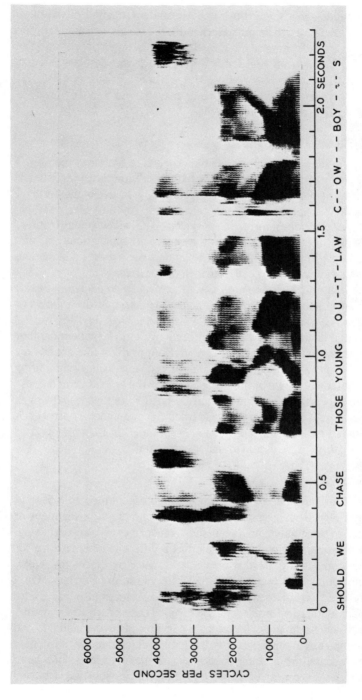

Fig. 9-6. Sound spectrogram of a sentence of speech spoken by a male voice (courtesy of Bell Telephone Laboratories, Inc.).

165

Kempelen's machine in boyhood, Alexander Graham Bell patented a procedure for producing speech in 1876. One important precursor of the modern digital devices for synthesizing speech was the analog Voder from Bell Laboratories which was demonstrated at the World Fairs in New York (1939) and in San Francisco (1940). It took a year to train operators to play the machine to produce simple, but recognizable speech.

DIGITAL SPEECH SYNTHESIS

Techniques for storing human speech in computer memories and playing back under specified, fixed conditions are widely used. Electrical machines of this type now talk to us in the form of language translators, talking calculators, spelling machines, as well as telephone information services. We shall be seeing (rather, hearing) a stream of other answer-back applications of this technique in the days ahead. This is not true synthesis of speech but only storage and recall from memory on demand.

It is interesting to note that to program a computer to talk, a model of speech production is necessary and that the models of Figs. 9-3, 9-4, and 9-5 have been applied in that way. Figure 9-7 shows a diagram of a digital synthesis system. A random number generator produces the digital equivalent of the s-like sounds for the unvoiced components. A counter produces pulses simulating the pulses of sound of the vocal cords for the voiced components. These are shaped by time varying digital filters simulating the ever changing resonances of the vocal tract. Special signals control each of these to form the digitized speech which is then changed to analog form in the digital to analog converter.

MUSIC

Musical sounds are extremely variable in their complexity and may range from a near sine wave form of a single instrument or voice to the highly complex mixed sound of a symphony orchestra. Each instrument and each voice has a different tonal texture for each note. Many musical instruments, such as the violin, viola, cello, or bass, produce their tones by vibration of strings. On a stretched string the overtones are all exact multiples of the fundamental, the lowest tone produced. These overtones may thus be called harmonics. If the string is bowed in the middle, odd harmonics are emphasized as the fundamental and odd harmonics have maximum amplitude there. Because the even harmonics have nodes in the center of the string, they will be subdued if bowed

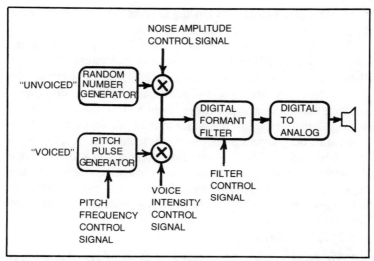

Fig. 9-7. A digital system for the synthesizing of speech. Note the similarity to the models of Figs. 9-3, 9-4, and 9-5.

there. The usual place for bowing is near one end of the strings which gives a better blend of even and odd harmonics. There is a problem with the seventh harmonic as it is not musically related to

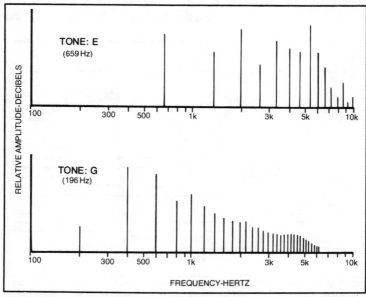

Fig. 9-8. Harmonic content of open strings of the violin. The lower tones have more closely packed harmonics.

the same musical family. By bowing 1/7th of the distance from one end, this harmonic is discriminated against.

The harmonic content of the E and G notes of a violin are displayed graphically in Fig. 9-8. Harmonic multiples of the higher E tone are spaced wider and hence have a "thinner" timbre. The lower frequency tone , on the other hand, has a closely spaced spectral distribution and a richer timbre. The small size of the violin relative to the low frequency of the G string means that the resonating body cannot produce a fundamental at as high a level as the higher harmonics. The harmonic content and spectral shape depend upon the shape and size of the resonating violin body, the type and condition of the wood, and even the varnish. Why there are so few superb violins among the many good ones is a problem that has not yet been solved completely.[61]

WIND INSTRUMENTS

We have considered resonances in the three dimensional room in detail in Chapter 6. In many musical instruments we must consider resonances in pipes or tubes on a primarily one dimension basis. Standing wave effects are dominant in pipes. If air is enclosed in a narrow pipe closed at both ends, the fundamental (twice the length of the pipe) and all its harmonics will be formed. Resonances will be formed in a pipe open at only one end at a frequency at which the pipe length is four times the wavelength and

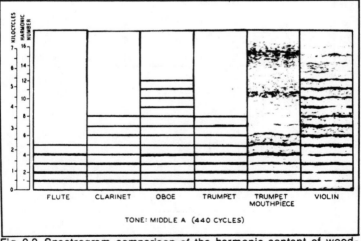

Fig. 9-9. Spectrogram comparison of the harmonic content of wood-wind instruments and the violin as middle A (440 Hz) is played. The differences displayed account for the differences in timbre of the different instruments (courtesy Bell Telephone Laboraboriies, Inc.).

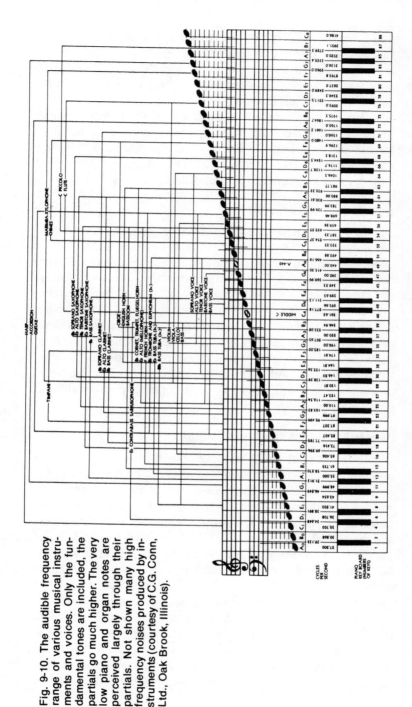

Fig. 9-10. The audible frequency range of various musical instruments and voices. Only the fundamental tones are included, the partials go much higher. The very low piano and organ notes are perceived largely through their partials. Not shown many high frequency noises produced by instruments (courtesy of C.G. Conn, Ltd., Oak Brook, Illinois).

odd harmonics are formed. Wind instruments form their sounds this way, the length of the column of air being continuously varied, as in the slide trombone, or in jumps as in the trumpet or French horn, or by opening or closing holes along its length as in the saxaphone, flute, and clarinet.

The harmonic content of several wind instruments is compared to that of the violin in the spectrograms of Fig. 9-9. Each instrument has its characteristic timbre determined by number and strength of its harmonics and by the formant shaping of the train of harmonics by structural resonances of the instrument.

NON-HARMONIC OVERTONES

In Chapter 2 the experience of Harvey Fletcher in synthesizing piano sounds[23] was mentioned. It was emphasized that piano strings are stiff strings and vibrate like a combination of solid rods and stretched strings. This fact means that the piano overtones are not strictly harmonic. Bells produce a wild mixture of overtones and the fundamental is not even graced with that name among specialists in that field. The overtones of drums are not harmonically related although they give a richness to the drum sound. Triangles and cymbals give such a mixture of overtones that they blend reasonably well with other instruments. Non harmonic overtones make the difference between organ and piano sounds and give variety to musical sounds.

DYNAMIC RANGE OF SPEECH AND MUSIC

In the concert hall a full symphony orchestra is capable of producing some very loud sounds when the score says so, but also some soft, delicate passages. Seated in the audience, one can fully appreciate this grand sweep of sound due to the great dynamic range of the human ear. The dynamic range between the loudest and the softest passage will be in the order of 60 to 70 dB. To be effective, the soft passages must still be above ambient background noise in the hall, hence the emphasis on adequate structural isolation to protect against traffic and other outside noises and precautions to assure that air handling equipment noise is low.

To those not present to hear the orchestra in the music hall reliance must be placed in AM or FM radio, television, magnetic recordings, or disc recordings. None of these is able to handle the full dynamic range of the orchestra but digital recording offers hope for the future. It boils down to noise limitations at the lower

Table 9-1. Power of Musical Sources.*

Instrument	Peak Power, Watts
Full orchestra	70
Large bass drum	25
Pipe organ	13
Snare drum	12
Cymbals	10
Trombone	6
Piano	0.4
Trumpet	0.3
Bass saxophone	0.3
Bass tuba	0.2
Double bass	0.16
Piccolo	0.08
Flute	0.06
Clarinet	0.05
French horn	0.05
Triangle	0.05

* Reference 62

extreme and distortion, and in the case of broadcasting, regulatory restrictions, at the upper end. Each type of channel has its own limitations and electronic or manual compression is used to

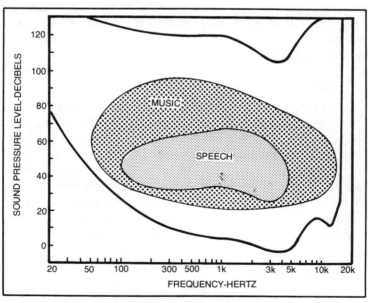

Fig. 9-11. Averaged and approximate sound pressure levels and frequency ranges of speech and music in relation to the extremes that can be handled by the human ear-brain mechanism. (Also, see Fig. 2-6).

squeeze the orchestra into the 45 to 50 dB range of high quality records and less than that in broadcasting.

POWER IN SPEECH AND MUSIC

In learning more about the various signals to be handled, the peak power of various sources must be considered. As for speech, the average power is only about 10 microwatts, but peaks might reach a milliwatt. Most of the power of speech is in the low frequencies, 80% below 500 Hz, yet there is very little power below 100 Hz. On the other hand, the small amount of power in the high frequencies determines the intelligibility of speech and thus is very important because that is where the consonants are. The peak power of various musical instruments is listed in Table 9-1.

FREQUENCY RANGE OF SPEECH AND MUSIC

It is instructive to compare the frequency range of the various musical instruments with that of speech and this is best done graphically. Figure 9-10 includes the ranges only of the fundamental and not of the harmonic tones of the instruments. The very low piano and organ notes, which are below the range of audibility of the ear, are perceived by their harmonics. Certain high frequency noise accompaniment of musical instruments is not included such as reed noise in woodwinds, bowing noise of strings, and key clicks and thumps of piano and percussion instruments.

AUDITORY AREA

The frequency range and the dynamic range of speech, music, and all other sounds place varied demands upon the human ear. Figure 9-11 shows the areas enclosed by the average intensity range and frequency range for speech and music. Only a portion of the auditory area, from Fig. 2-6, is used. The auditory area, bounded by the threshold of hearing and the threshold of pain, encompasses all of the auditory stimuli to which man responds. It is apparent that even speech and music, with all their extreme variations, do not exhaust the capabilities of the human ear.

Absorption of Sound 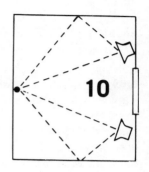 10

The law of the conservation of energy states that energy can neither be created nor destroyed but that it can be changed from one form to another. If we have some sound energy in a room which we wish to get rid of, how can it be done? Sound is the vibratory energy of air particles and it can be dissipated in the form of heat. If it takes the sound energy of a million people talking to brew a cup of tea we must give up any idea of heating our home with sound from the hi-fi. However, a start might be made if the power amplifier and loudspeakers are sufficiently low in efficiency, but this is electrical, not sound energy.

POROUS MATERIALS

The key word in this discussion of porous sound absorbers is *interstices*. It means simply the space between two things and the way it rolls around on the tongue is so much more satisfying than the harsh synonyms *crevice, crack*, or *chink*. If a sound wave strikes a wad of cotton batting, the sound energy sets the cotton fibers to vibrating. The fiber amplitude will never be as great as the air particle amplitude of the sound wave because of frictional resistance. Some sound energy is changed to frictional heat as the cotton fibers are set in motion, restricted as this motion is. The sound penetrates more and more into the interstices of the cotton, losing more and more energy as more and more fibers are vibrated. Cotton is an excellent sound absorber which I have specified in

173

studio treatment in parts of Africa because it was plentiful and cheap and imported materials were out of the question.

There are proprietary porous materials available for sound absorption. These are commonly of cellulose or mineral fiber. Their absorption effectiveness depends on the density of packing of the fibers, thickness of the resulting layer, and the method of mounting.

ABSORPTION COEFFICIENT

The sound falling on a surface is either reflected, absorbed, or transmitted. In Fig. 10-1 a sound absorbing material is applied to a brick or concrete block wall. A sound ray S strikes the absorbent at a given angle. A very small amount of sound energy, a, is reflected from the surface of the absorbent. A greater amount, b, is reflected from the hard wall surface because a greater difference in density is encountered. Refraction or bending of the sound ray takes place as the sound wave front enters the absorbent, as it enters the masonry wall, and also as it leaves the wall on the far side, all this bending according to the principles illustrated in Fig. 4-6. Sound energy is lost in the interstices of both the absorbent and the wall. All the energy in S is accounted for by the weakened rays a, b, and c which will bounce around in other encounters of the S kind and the heat losses d and e.

The absorption coefficient is our measure of the efficiency of the absorbent in absorbing sound. If 55% of the incident sound energy is absorbed the absorption coefficient is said to be 0.55. One square foot of this material gives o.55 absorption units (sabins), as comparing it to a perfect absorber having an absorption coefficient of 1.0.

The absorption coefficient of a given material varies with the angle of incidence. As used in this book, the absorption coefficients are taken as the average for all angles of incidence. The average absorption coefficient also varies with frequency and in tables to follow coefficients are given for the following frequencies at octave spacings: 125, 250, 500, 1000, 2000, and 4000 Hz. Absorption coefficients also vary with type of mounting, thickness of material, etc., which will be examined in detail.

ABSORPTION COEFFICIENT AND SOUND LEVEL

If one is considering the relative strengths of several reflected components in an auditorium, for instance, it is necessary to know the relationship between absorption coefficients and the decrease

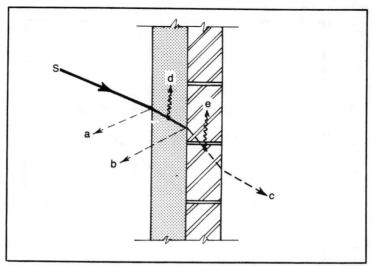

Fig. 10-1. A sound ray impinging on an acoustical material on a masonry wall undergoes reflection from two surfaces, absorption in two materials, and refraction. In this chapter the absorbed component is of chief interest.

of sound pressure level in dB upon reflection which is given by the expression:

$$D = 10 \log_{10} \frac{1}{1 - a} \qquad (10-1)$$

where,   D = decrease in sound pressure level, dB
         a = energy absorption coefficient.

If a material has an absorption coefficient of a = 0.9 at a given angle and frequency the reflected sound pressure level is $10 \log_{10} (1/0.1) = 10$ dB lower than the incident sound pressure level. If a = 0.50 the decrease in sound pressure level is only $10 \log_{10} (1/0.5) = 3$ dB. Figure 10-2 is a plot of absorption coefficient a and the decrease in sound pressure level in dB computed from Equation 10-1.

EFFECT OF THICKNESS

It is logical to expect greater sound absorption from thicker materials, but this logic holds primarily for the lower frequencies. Figure 10-3 shows the effect of varying absorbent thickness where the absorbent is mounted directly on a solid surface (mounting #4). In Fig. 10-3 there is little difference above 500 Hz in increasing the

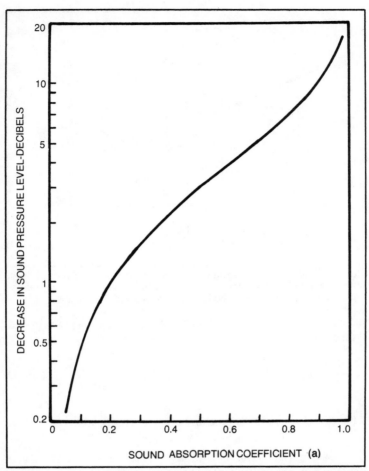

Fig. 10-2. A ray of sound reflected from an acoustical material has a lower sound pressure level than the incident ray. This curve shows the decrease in sound pressure level in dB in relation to the absorption coefficient of the material.

absorbent thickness from 2″ to 4″ but there is considerable improvement below 500 Hz as thickness is increased. There is also a proportionally greater gain in overall absorption in a 1″ increase of thickness in going from 1″ to 2″ than going from 2″ to 3″ or 3″ to 4″. A 4″ thickness of glass fiber material of 3 lb/cu ft density has essentially perfect absorption over the 125 Hz - 4 kHz region.

EFFECT OF AIR SPACE

Low frequency absorption may also be improved by spacing the absorbent out from the wall. This is an inexpensive way to get improved performance - within limits. Figure 10-4 shows the effect on absorption coefficient of furring 1″ glass fiber wall board out from a solid wall. Spacing 1″ material out 3″ makes its absorption approach that of the 2″ material of Fig. 10-3 mounted directly on the wall.

EFFECT OF DENSITY

Glass fiber and other materials come in various densities from the flimsy thermal insulation batts to the semirigid and rigid boards used widely in industry. All of these have their proper place in acoustical treatment of spaces, but our question right now is, "What effect does density, the packing of the fibers, have on surface absorption coefficient?" In other words, is the sound able to penetrate the interstices of the high density, harder surface material as well as one of the flimsy kind? The answer appears in Fig. 10-5 which shows relatively little difference in absorption coeffi-

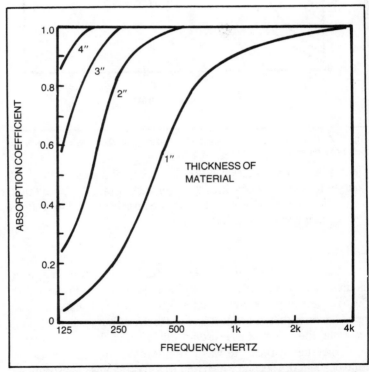

Fig. 10-3. The thickness of glass fiber sound absorbing material determines the low frequency absorption. Density, 3 lbs/cu ft, material mounted directly on hard surface.

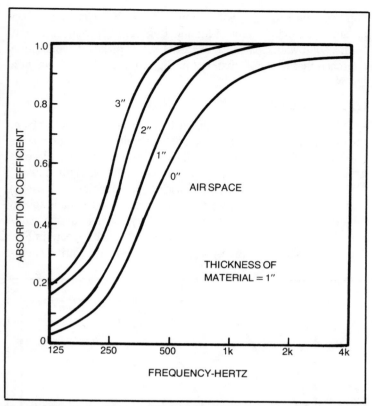

Fig. 10-4. The low frequency absorption of 1 inch glass fiber board is improved materially by spacing it out from the solid wall.

cient as the density is varied over a range of almost 4 to 1. Of course, for very low densities the fibers are so widely spaced that absorption suffers. For extremely dense boards the surface reflection is high and sound penetration low.

CARPET

Carpet commonly dominates the acoustical picture in spaces as diverse as living rooms, recording studios, and churches. It is the one amenity the owner often specifies in advance and the reason is more often comfort and appearance, not acoustical. For example, the owner of a recording studio with a floor area of 1000 sq ft specified carpet. He was also interested in a reverberation time of about 0.5 second which requires 1060 sabins of absorption. At the higher audio frequencies a heavy carpet and pad with an absorption coefficient of around 0.6 gives 600 sabins absorption at

4 kHz or 57% of the required absorption for the entire room before the absorption needs of walls and ceiling are even considered. The acoustical design is almost frozen before it is started.

There is another, more serious problem. This high absorbence of carpet is only at the higher audio frequencies. Carpet having an absorption coefficient of 0.60 at 4 kHz offers only 0.05 at 125 Hz. In other words, the 1000 sq ft of carpet introduces 600 sabins absorption at 4 kHz but only 50 sabins at 125 Hz! This is the first major problem encountered in many acoustical treatment jobs. The unbalanced absorption of carpet can be compensated in other ways, principally with resonant type absorbers.

To compound the problem of unbalanced absorption of carpet, dependable absorption coefficients are hard to come by. A bewildering assortment of types of carpet and variables in underlay add to the uncertainty. Unfortunately, reverberation chamber

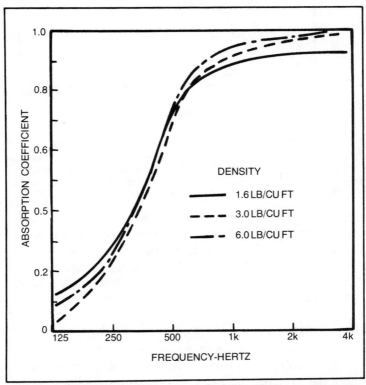

Fig. 10-5. Density of the glass fiber sound absorbing material has relatively little effect on absorption in the range 1.6 to 6 lb/cu ft. Material mounted directly on solid wall.

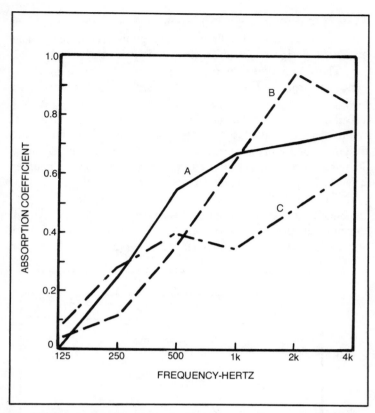

Fig. 10-6. A comparison of the sound absorption characteristics of three different types of carpet: (A) Wilton, pile height 0.29″, 92.6 oz/sq yd, (B) Velvet, latex backed, pile height 0.25″, 76.2 oz/sq yd, and (C) the same Velvet without latex backing, 37.3 oz/sq yd, all with 40 oz hairfelt underlay (after Harris, Ref. 63).

measurements of random incidence absorption coefficients for specific samples of carpet are involved and expensive and, generally, unavailable to the acoustical designer. Therefore, all we can do is to be informed on the factors affecting the absorption of carpets and make judgements on what coefficients to use for the specific carpet at hand.

Effect of Carpet Type

What variations in sound absorption should one expect between types of carpet? Figure 10-6 shows the difference between a heavy Wilton carpet and a velvet carpet with and without a latex back. The latex backing seems to increase absorption materially above 500 Hz and to decrease it a modest amount below 500 Hz.

Effect of Carpet Underlay

Hairfelt was formerly used almost exclusively as the padding under the carpet. It is interesting that in the early days of this century hairfelt was used for general acoustical treatment until displaced by numerous proprietary materials. Today foam rubber, sponge rubber, felts, polyurethane, or combinations have replaced hairfelt. Foam rubber is made by whipping a latex water dispersion, adding a gelling agent, and pouring into molds. The result is always open celled. Sponge rubber, on the other hand, formed by chemically generated gas bubbles, may yield either open or closed cells. Open cells provide the interstices required for good sound absorption while closed cells do not.

The influence of underlay on carpet absorption is very great. In Fig. 10-7 are shown chamber measurements of absorption coefficients of a single Axminster type of carpet with different underlay conditions. Graphs A and C show the effect of hairfelt of 80 and 40

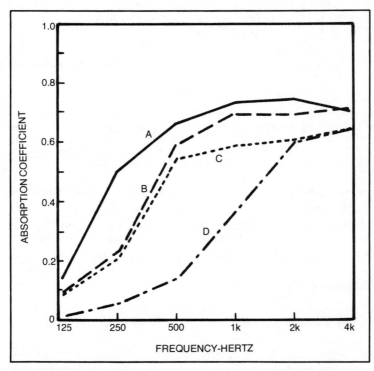

Fig. 10-7. Sound absorption characteristics of the same Axminster carpet with different underlay: (A) 80 oz hairfelt, (B) hairfelt and foam, (C) 40 oz hairfelt, and (D) no underlay, on bare concrete (after Harris, Ref. 63).

181

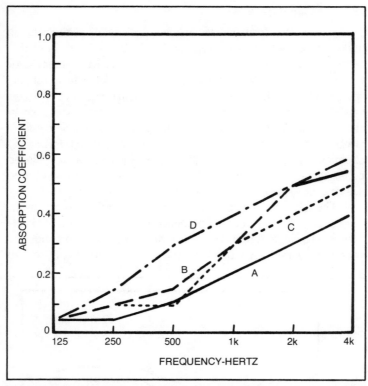

Fig. 10-8. Carpet absorption coefficients from commonly used table: (A) 1/8″ pile height, (B) 1/4″ pile height. (C) 3/16″ combined pile and foam, and (D) 5/16″ combined pile and foam. These graphs are to be compared to those of Figs. 10-6 and 10-7 (from Ref. 64, reproduced in the appendix).

oz/sq yd weight. Graph B shows an intermediate combination of hairfelt and foam. While these three graphs differ considerably, they all stand in stark contrast to graph D for the carpet layed directly on bare concrete. Conclusion: the padding underneath the carpet contributes markedly to overall carpet absorption.

STANDARD COEFFICIENTS

The absorption coefficients plotted in Figs. 10-6 and 10-7 are taken from Harris' 1957 paper,[63] perhaps the most exhaustive study of carpet characteristics available. In general his coefficients are higher than those currently offered in published tables. In Fig. 10-8 the coefficients listed in a widely used publication[64] and included in the appendix are plotted for comparison with Figs 10-6 and 10-7. Carpets vary widely which may account for some of the great variability with which any designer of acoustical systems is confronted.

FOAMS

A new type of sound absorber has appeared during the last decade or so, foams. Flexible polyurethane foams are widely used in noise quieting of automobiles, machinery, aircraft and in various industrial applications. After a slow start foams are finding some application as sound absorbers in architectural applications, including recording studios and home listening rooms. Figure 10-9 is a photograph of Sonex, a foam product contoured to simulate the wedges used in anechoic rooms. These are shaped in male and female molds and come in meshed pairs. This material may be cemented or stapled to the surface to be treated. The absorption coefficients of Sonex of different thicknesses are shown in Fig. 10-10.

DRAPES

Drapes must also be considered a porous type of sound absorber because air can flow through the fabric under pressure. Variables affecting absorbency include weight of material, degree of drape, and distance from the wall. Data are scarce but Fig. 10-11

Fig. 10-9. Sonex contoured acoustical foam simulating anechoic wedges. This is an open cell type of foam which offers better sound absorption than closed cell type.

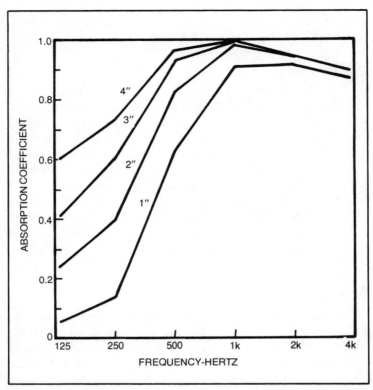

Fig. 10-10. Absorption coefficients of Sonex contoured foam of various thicknesses (from Ref. 65).

compares the absorption of 10, 14, and 18 oz/sq yd velour hung straight and presumably at some distance from any wall. One intuitively expects greater absorption with heavier material. However, the greater absorption in going from 14 to 18 oz/sq yd than going from 10 to 14 oz/sq yd is difficult to explain. The effect, whatever it is, is concentrated in the 500 to 1 kHz region.

The amount of fullness of the drape has a great effect as shown in Fig. 10-12. The "draped to ⅞ area" means that the entire 8/8 area is drawn in only slightly (⅛th) from the flat condition. The deeper the drape fold, the greater the absorption.

The distance a drape is hung from a reflecting surface can have a great effect on its absorption efficiency. This is best explained by Fig. 10-13. In Fig. 10-13A a drape or other porous material is hung parallel to a solid wall and the distance d between the two is varied. The frequency of the sound impinging on the porous material is held constant at 1000 Hz. If the sound absorption provided by the

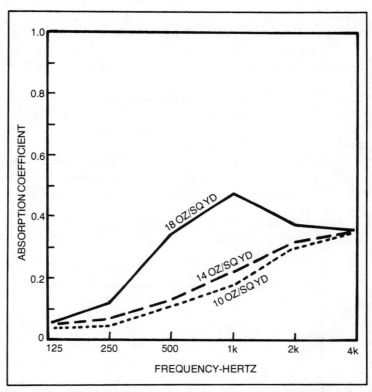

Fig. 10-11. The sound absorption of velour hung straight for three different weights of fabric (after Beranek, Ref. 68).

porous material is measured we find that it varies greatly as the distance d from the wall is changed. Looking at the situation closely reveals that the wavelength of the sound is related to maxima and minima of absorption. The wavelength of sound is the speed of sound divided by frequency which, in the case of 1000 Hz, is $1130/1000 = 1.13$ ft or about 13.6 inches. A quarter wavelength is 3.4 inches and half wavelength is 6.8 inches. We note that there are absorption peaks at ¼ wavelength and, if we carry it further than indicated in Fig. 10-13A, at each odd multiple of quarter wavelengths. Absorption minima occur at even multiples of quarter wavelengths.

These effects are explained by reflections of the sound from the solid wall. At the wall surface, pressure will be highest, but air particle velocity zero because the sound waves cannot supply enough energy to shake the wall. At a quarter wavelength from the wall, however, pressure is zero and air particle velocity is maximum. By placing the porous material, such as a drape, a quarter wavelength from the wall it will have maximum absorbing effect because the particle velocity is maximum. At half wavelength particle velocity is at a minimum, hence absorption is minimum.

In Fig. 10-13B the spacing of the drape from the wall is held constant at 12″ as the absorption is measured at different frequencies. The same variation of absorption is observed, maximum when the spacing is at odd quarter wavelengths and minimum at even quarter wavelengths from the wall. At this particular spacing of 12″ or 1 foot a wavelength of spacing occurs at 1130/1 = 1130 Hz and a quarter wavelength at 276 Hz and a half wavelength at 565 Hz, etc.

In Fig. 10-14 are shown actual reverberation chamber measurements of the absorption of 19 oz/sq yd velour. The solid graph,

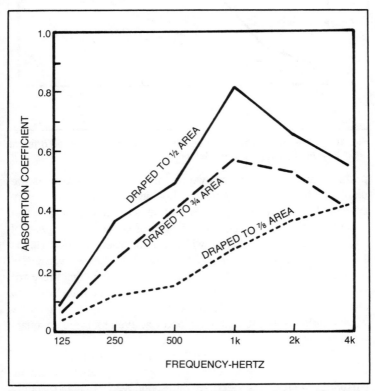

Fig. 10-12. The effect of draping on the sound absorption of drapes. "Draped to ½ area" means that folds are introduced until the resulting drape area is half that of the straight fabric (after Mankovsky, Ref. 69).

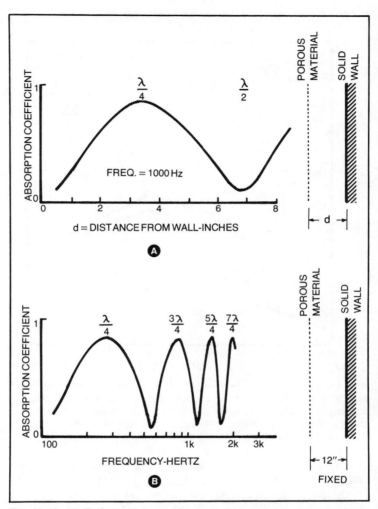

Fig. 10-13. (A) The sound absorption of a porous material such as a drape varies with the distance from a hard wall. Maximum absorption is achieved when the drape is one quarter wavelength from the wall, minimum at half wavelength. (B) The sound absorption of a porous material hung at a fixed distance from a wall will show maxima at spacings of quarter wavelength and odd multiples of quarter wavelength as the frequency is varied.

presumably, is for a drape well removed from all walls. The other graphs, very close together, are for the same material spaced 10 cm (about 4″) and 20 cm (about 8″) from the wall. The 10 cm distance is one wavelength at 3444 Hz, the 20 cm distance at 1722 Hz. The odd multiples of both the 10 cm and the 20 cm quarter wavelengths are spotted on the upper part of Fig. 10-14.

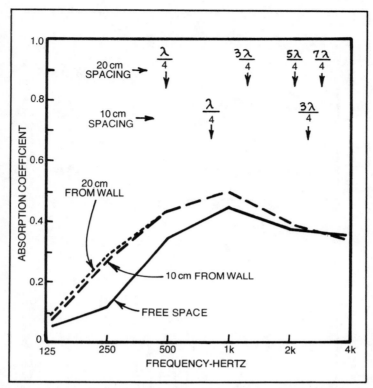

Fig. 10-14. Actual measurements of sound absorption coefficients of a velvet material(19 oz/sq yd)in free space and 10 cm and 20 cm from a solid wall. The points at which increase in absorption due to wall reflection is to be expected are indicated (after Mankovsky, Ref. 69).

The absorption of the velour is greater when spaced from the wall and the effect is greatest in the 250 Hz - 1 kHz region. At 125 Hz the 10 and 20 cm spacing adds practically nothing to the drape absorption because at 125 Hz the quarter wavelength spacing is 2.26 ft. We must remember that when we talk quarter wavelengths we infer sine waves. Absorption measurements are invariably made with bands of random noise. Hence we must expect the wiggles of Fig. 10-13B to be averaged out by the use of such bands.

SOUND ABSORPTION BY PEOPLE

People absorb sound too. Just how much of this is due to absorption by flesh and how much by hair and clothing has yet to be reported and such X-rated research projects to get such answers appear unlikely. The important thing is that the people making up

**Table 10-1. Sound Absorption
by People. (Sabins Per Person)**

	Frequency, Hz					
	125	250	500	1k	2k	4k
College students informally dressed seated in tablet arm chairs (Ref. 66)	—	2.5	2.9	5.0	5.2	5.0
Audience seated, depending on spacing and upholstery of seats (Ref. 64)	2.5– 4.0	3.5– 5.0	4.0– 5.5	4.5– 6.5	5.0– 7.0	4.5– 7.0

an audience account for a significant part of the sound absorption of the room. It also makes an acoustical difference whether one or ten people are in a small monitoring room. The problem is how to rate human absorption and how to involve it in calculations. The usual method of multiplying a human absorption coefficient by the area of a human has its problems. The easy way is to determine the absorption units (sabins) a human presents at each frequency and add them to the sabins of the carpet, drapes, and other absorbers in the room at each frequency. Table 10-1 lists the absorption of informally dressed college students in a classroom along with a range of absorption for more formally dressed people in auditorium environment.

We see that for 1 kHz and higher, the absorption offered by college students in informal attire in the spartan furnishings of a classroom falls at the lower edge of the range of a more normal audience. The low frequency absorption of the students, however, is considerably lower than the more formally dressed people.

SOUND ABSORPTION IN AIR

For frequencies 1 kHz and above and for very large auditoriums the absorption of sound by the air in the space becomes

Table 10-2 Sound Absorption in Air 50% Relative Humidity.

Frequency	Absorption (Sabins Per 1,000 cu ft)
1000 Hz	0.9
2000	2.3
4000	7.2
(From Reference 64)	

important. For example, a church seating 2000 has a volume of about 500,000 cu ft. Referring to Table 10-2 we note that for 50% relative humidity the absorption is 7.2 sabins per 1000 cu ft or a total of (500) (7.2) = 3600 sabins at 4 kHz. This is equivalent to 3600 sq ft of perfect absorber.

This could be 20-25% of the total absorption in the space and there is nothing that can be done about it other than to take it into consideration and taking consolation in at least knowing why the treble reverberation time falls off so much!

LOW FREQUENCY ABSORPTION

The absorption of sound at the lower audible frequencies may be achieved by (a) porous absorbers or (b) by resonant absorbers. Glass fiber and acoustical tiles are common forms of porous absorption in which the sound energy is dissipated as heat in the interstices of the fibers. The absorption of commercial forms of glass fiber and other fibrous absorbers at low audio frequencies, however, is quite poor. To absorb well the thickness of the porous material must be comparable to the wavelength of the sound. At 100 Hz the wavelength is 11.3 feet and using any porous absorber approaching this thickness would be somewhat impractical. So, we turn our attention to the resonant type of absorber to obtain good absorption at low frequencies.

DIAPHRAGMATIC ABSORBERS

We leave the porous type of absorber and their precious interstices and consider several resonant types. The simplest is the diaphragm vibrating in response to sound and absorbing some of that sound because of frictional heat losses in the fibers as it flexes. A piece of ¼" plywood is an excellent example. Let us assume that it is spaced out from the wall on 2 x 4s which gives close to 3-¾" air space behind. The frequency of resonance of this structure may be calculated from the expression:

$$f_o = \frac{170}{\sqrt{(m)(d)}} \qquad (10\text{-}2)$$

where,

f_o = frequency of resonance, Hz
m = surface density of the panel, lb/sq ft of panel surface
d = depth of air space, inches.

The surface density of ¼″ plywood, 0.74 lb/sq ft, can be measured or found in the books. Substituting in equation 10-2 we get:

$$f_o = \frac{170}{\sqrt{(0.74)\,(3.75)}}$$

$$f_o = 102 \text{ Hz}$$

Some of the great chamber music rooms owe their acoustical excellence to the low frequency absorption offered by extensive paneled walls.

Plywood panels still give good low frequency absorption even if wrapped into a semi-cylindrical surface with the added advantage of excellent diffusing properties.[67] Plywood or tongue and groove flooring or subflooring vibrates as a diaphragm and contributes to low frequency absorption. Drywall construction on walls and ceiling does the same thing. All such components of absorption must be included in the acoustical design of a room, large or small.

HELMHOLTZ RESONATORS

The Helmholtz type of resonator is widely used to achieve adequate absorption at lower audio frequencies. There is nothing particularly mysterious about them, in fact they pop up in various forms in everyday life. Blowing across the mouth of any bottle or jug produces a tone at its natural frequency of resonance. The air in the cavity is springy and the mass of the air in the neck of the jug reacts with this springiness to form a resonating system, much as a weight on a spring vibrating at its natural period. Change the volume of the air cavity, the length or diameter of the neck and you change the frequency of resonance. Such a Helmholtz resonator has some very interesting characteristics. For instance, sound is absorbed at the frequency of resonance and at nearby frequencies. The width of this absorption band depends upon the friction of the system. A glass jug offers little friction to the vibrating air and would have a very narrow absorption band. Adding a bit of gauze across the mouth of the jug or stuffing a wisp of cotton into the neck, the amplitude of vibration is reduced and the width of the absorption band is increased.

The sound impinging on a Helmholtz resonator which is not absorbed is reradiated. As the sound is reradiated from the resonator opening, it tends to be radiated in all directions. This means that unabsorbed energy is diffused and diffusion of sound is a very desirable thing in a studio or listening room.

Bottles and jugs are not appropriate forms of Helmholtz resonators to apply the resonance principle in studios. An interesting experiment conducted many years ago at Riverbank Acoustical Laboratories bears this out.[70] To demonstrate the effectiveness of a continuously swept narrow band technique of measuring sound absorption coefficients the idea was conceived to measure the absorption of Coca Cola bottles. A tight array of 1152 empty 10 ounce bottles was arranged in a standard 8 x 9 foot space on the concrete floor of the reverberation chamber. It was determined that a single, well isolated bottle has an absorption of 5.9 sabins at its resonance frequency of 185 Hz, but with a bandwidth (between −3dB points) of only 0.67 Hz! Absorption of 5.9 sabins is an astounding amount of absorption for a Coke bottle! This is about what a person, normally clothed, would absorb at 1000 Hz, or what 5.9 sq ft of glass fiber (2″ thick, 3 lb/cu ft density) would absorb at midband. The sharpness of this absorption characteristic is even more amazing. This would correspond to a Q of 185/0.67 = 276! As interesting as these data are they tell us that leaving an empty Coke bottle in a studio will not devastate the acoustics of the room, but it might have a tiny effect at 185 Hz.

Well, if bottles are not a suitable form of Helmholtz resonators for a studio, what is? In Fig. 10-15 we have conveniently idealized a square bottle with a tubular neck on it. Of course, this bottle alone would produce its characteristic tone if one were to blow across the opening. Stacking these bottles does not detract from the resonator action, but rather enhances it. It is a small step to a box of length L, width W, and depth H which has a lid of thickness equal to the length of necks of the bottles. In this lid are drilled holes having the same diameter as the holes in the neck. It is just a bit harder to realize that partitions between each segment may be removed without greatly affecting the Helmholtz action. In this way a Helmholtz resonator of the perforated face type can be related to funny shaped bottles giving us something of a physical picture of how perforated face resonators perform.

In a similar way Fig. 10-16 illustrates another funny bottle with an elongated slit neck. These, too, can be stacked, even in multiple rows and we see that it is but a short step to a slot type resonator. Here also we can dispense with separating walls in the air cavity without destroying the resonator action. A word of caution is in order. Subdividing the air space can improve the action of perforated face or slit resonators but only because this reduces spurious, unwanted modes of vibration being set up within the air cavity.

192

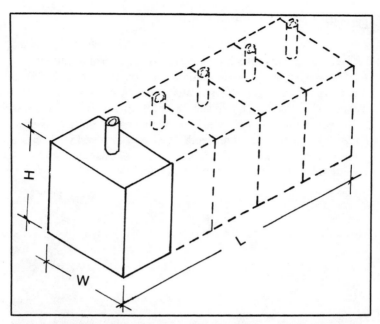

Fig. 10-15. Development of a perforated face Helmholtz resonator from a single rectangular bottle resonator.

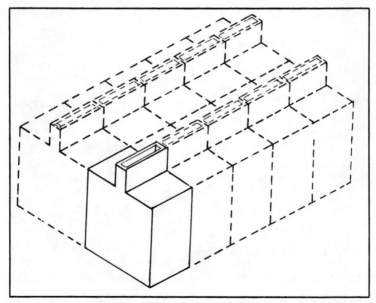

Fig. 10-16. Development of a slot type Helmholtz resonator from a single rectangular bottle resonator.

193

EXPERIMENTS

It was decided to make some measurements on various forms of Helmholtz resonators applicable to acoustical treatment of studios and listening rooms. Rather than construct many boxes, a fixed box with changeable faces was selected. An old loudspeaker cabinet of ⅝" particle board and 23" x 30" inside dimensions was cut down to a depth of 7-⅞". That depth was carefully chosen to coincide with the 200 mm depth used by Mankovsky[69] because measured absorption coefficients for boxes of this depth and several different facings are included in his book.

Fig. 10-17. Experimental Helmholtz resonator with means for changing the type of cover which was used for program of measurements. The cover shown has a perforation percentage of 0.42% (perf-B of Table 10-3).

Fig. 10-18. Arrangement for outdoor measurement of sound pressure inside a Helmholtz resonator as a function of frequency. The B&K Type 4165 calibrated ½" condenser microphone is held in place by a crossbar arrangement.

The wing nut arrangement for changing faces is shown in Fig. 10-17. The back was also secured with wood screws to facilitate removal and support for the calibrated ½" condenser microphone of the Brüel & Kjaer 2215 sound level meter installed as in Fig. 10-18. Tightness between the changeable faces and the box was assured by a strip of sponge rubber. Potential cracks and a few actual holes were sealed with the application of a non-hardening sealant commonly used around bathtubs and in shower stalls. A

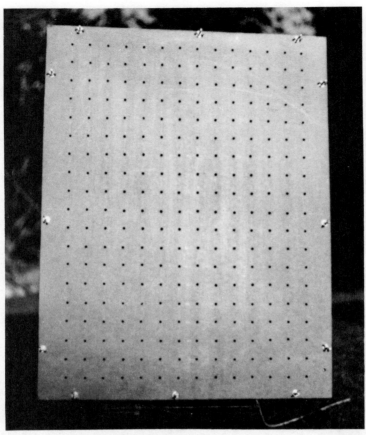

Fig. 10-19. Helmholtz perforated type resonator with a cover having 1.05% of its effective area in holes (perf-A of Table 10-3).

Table 10-3. Resonator Specifications.

	Perf-A	Perf-B	Perf-C	Slat
	Resonator Type			

	Perf-A	Perf-B	Perf-C	Slat
Mankovsky[1]				
Cover thickness	4 mm	4 mm	4 mm	
Hole diameter	5 mm	5 mm	5 mm	
Hole spacing	35 mm	65 mm	100 mm	
Depth, air space	200 mm	200 mm	200 mm	
Perforation %	1.60%	0.46%	0.196%	
Calculated f_o	169 Hz	91 Hz	59 Hz	
Ours				
Cover thickness	3/16''	3/16''	3/16''	Slat thickness 3/4''
Hole diameter	3/16''	3/16''	3/16''	Slot width 1/16''
Hole spacing	1-5/8''	2-9/16''	3-15/16''	Slat width 3-9/16''
Depth, air space	7-7/8''	7-7/8''	7-7/8''	Air space 7-7/8''
Perforation %	1.05%	0.42%	0.18%	
Calculated f_o	126 Hz	80 Hz	52 Hz	107 Hz

Fig. 10-20. Helmholtz perforated type resonator with a cover having 0.18% of its effective area in holes (perf-C, Table 10-3).

small sheet of sponge rubber was tightly wrapped around the microphone cable and wedged into the hole in the back from which the cable emerges.

Three perforated faces were intentionally dimensioned to approximate those used by Mankovsky as shown in Table 10-3. These three covers were drilled for 1.05% (Perf-A, Fig. 10-19), 0.42% (Perf-B, Fig. 10-17), and 0.18% (Perf-C, Fig. 10-20) perforation. Our Perf-B and Perf-C covers come reasonably close to those of Mankovsky, but Perf-A fell victim to illegible handwriting. The hole spacing should have been 1-⅜″. Admittedly, a new panel

197

should have been made but this was discouraged by the fact that even at 1⅝" spacing there are 285 holes in the panel!

The perforation percentage is readily calculated on a single element of the repeat pattern. In Fig. 10-21 the perforation percentage is simply the area of one hole (four quarter holes) divided by the area between the holes multiplied by 100. Warning: if you dive into the literature on this subject be alert to confusion between "perforation percentage" and "perforation ratio" or "perforation fraction". If the fraction of hole area to panel area is not multiplied by 100, it is not properly labeled as percentage.

In addition to the perforated covers a single slat type cover was fitted to the box, shown in Fig. 10-22.

EXPERIMENTAL METHOD

The experimental method employed was a very simple one suggested by Goyer and reported by Beranek.[71] If an anechoic chamber or reverberation chamber is not available, the sound pressure inside a Helmholtz resonator can be compared to that outside the resonator to determine the frequency to which the resonator is tuned. A schematic diagram of instrumentation is

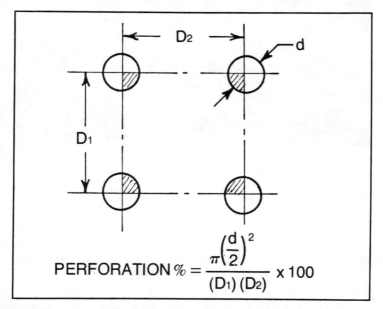

$$\text{PERFORATION} \% = \frac{\pi\left(\frac{d}{2}\right)^2}{(D_1)(D_2)} \times 100$$

Fig. 10-21. The perforation percentage is the ratio of total hole area to total effective area expressed as a percentage. In a repetitive pattern this is readily calculated from one unit of the repeated pattern.

Fig. 10-22. Arrangement of loudspeaker and experimental Helmholtz resonator with a slat type cover. Located on a hillside, bothersome reflections are minimized.

shown in Fig. 10-23. The physical layout, shown in Fig. 10-24, is located outdoors on a hillside location. It is a very rough estimation that a reflection from the Bank of America building a mile or so away would be down about 120 dB. Would that were true of reflections from the nearby Jacaranda tree! By keeping the level up around 110-120 dB, environmental noise effects were minimized even on low frequency octave bands. For sounds of this level ear protectors as well as sympathetic neighbors are definitely recommended (Fig. 10-24).

The method used involves comparing the sound pressure within the resonator to that outside the resonator as the frequency is varied in a standard way. Ideally, a dual channel measuring system would enable recording of both inside and outside pres-

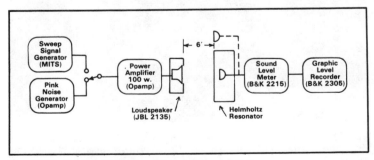

Fig. 10-23. Schematic diagram of measuring system.

sures simultaneously. The procedure used was to make an outside pressure run (Fig. 10-25) at the beginning and end of a series of measurements inside the box for various covers and glass fiber damping. To obtain the outside pressure the resonator was moved to one side and the measuring microphone placed exactly where the face of the resonator had been. Obviously, the very presence of the resonator itself as well as reflections from the ground and surrounding objects affect the outside pressure somewhat. However, every effort was made to make all measurements under conditions as nearly identical as possible so that comparisons between the different resonator facings could be made with reasonable confidence. In other words, the procedure is not ideal but suitable for comparative measurements on a low budget.

SINE SWEEP TESTS

Figure 10-26 shows tracings from graphic level recorder tapes of sine sweep tests. At some frequencies the sound pressure inside the resonator (solid line) is above the sound pressure prevailing at the outside face of the resonator (broken line). This is what we would expect near resonance. At other frequencies the inside pressure falls below the outside pressure. At frequencies well removed from resonance it is logical to expect the box to shield the

Fig. 10-24. Instrument arrangement of the outdoor "anechoic" facility. Ear protectors are required because of the 110-120 dB levels used to minimize the effects of environmental noise.

Fig. 10-25. To compare the sound pressure within a Helmholtz resonator to that outside, the resonator is moved to one side and the measuring microphone placed where the face of the resonator was located.

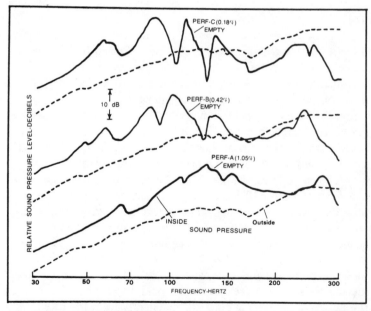

Fig. 10-26. Sine sweep comparison of sound pressure inside the Helmholtz resonator (solid line) with the corresponding sound pressure at the face of the resonator (broken line). Results for three perforation percentages are shown. No glass fiber is in the resonators.

201

microphone inside. The sine sweep, in a way, reveals too much information in the form of up and down fluctuations resulting from complex spurious resonances of box panels and modal effects within the box. However, with no glass fiber in the box (Fig. 10-26) a general progression of peaks from lower to higher frequencies as perforation percentage is increased can be noted. Not shown are sine sweep records taken with glass fiber in the box which are very difficult to interpret.

Similar sine sweep results for the slat facing are shown in Fig. 10-27. Pressing the glass fiber against the back of the slat facing reduces the resonance effect, although the same glass fiber at the rear of the box has relatively little effect other than shifting the peak to a somewhat lower frequency as expected.

ONE-THIRD OCTAVE SWEEPS

A magnetic recording was made of the output of a General Radio Type 382 random noise generator fed through a General Radio Type 1564 sound analyzer as it swept a ⅓ octave band. The playback of this recording was then used to drive the power amplifier and loudspeaker in the same manner as the swept sine test.

The ⅓ octave sweep recordings of Fig. 10-28 are considerably less erratic than the sine sweeps of Figs. 10-26 and 10-27 as would be expected from the averaging effect of the wider band. Measurements using ⅓ octave bands of noise make sense in that, over much of the audible spectrum, the human ear analyzes sound in critical bands roughly approximated by ⅓ octaves. However, in the 50-250 Hz region in which Helmholtz resonators are of greatest practical value, the critical bands of the ear are about 100 Hz wide which is better approximated by an octave bandwidth.

The recordings of Figs. 10-28 A, B, and C show nice resonance peaks of 10 or 12 dB. One mystery is the lack of agreement of peak frequency between the ⅓ octave sweeps (Fig. 10-28) and the sine sweeps (Figs. 10-26 and 10-27). Operator error is strongly suspected. One difference in procedure is that in the ⅓ octave sweep runs the Helmholtz box was closer to the loudspeaker (about 3 feet) and the runs were made indoors. Reverberation measurements (to be treated later) would, however, indicate that the box was performing quite independently from the room, yet standing waves could have had their usual unexpected and devastating effect. There is no need to be too concerned by such discrepencies as we move to more significant matters.

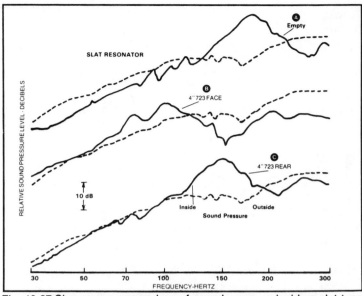

Fig. 10-27. Sine sweep comparison of sound pressure inside a slat type Helmholtz resonator:(A)no glass fiber inside,(B) 4" of Type 723 Fiberglas against the face,and (C) 4" of Type 723 Fiberglas against the rear of the box.

OCTAVE BAND MEASUREMENTS

For octave band measurements wideband pink noise was emitted from the loudspeaker at high level. The sound was measured both outside and inside the box with the aid of octave band filters which are an integral part of the B&K 2215 sound level meter. The sound level meter is acting normally in every respect, even though its microphone is on a long extension cable. Everything is held constant during the runs except for changing of resonator faces and internal glass fiber.

These curves beautifully depict resonance rise of inside pressure relative to outside pressure due to Helmholtz action. Each experimental point, shown by a circle, was obtained by subtracting from the dB sound pressure level the outside sound pressure level. The dB difference, plotted against frequency, shows the resonance effect, but in subdued form because of the logarithmic compression. To dramatize the peaking effect the dB difference was changed to linear sound pressure difference by solving $10 \ 20^{dB/20}$ for each frequency. The peak sound pressure difference of about 4 in Fig. 10-29A corresponds to a level rise of about 12 dB which is about that observed in Figs. 10-26, 10-27, and 10-28. In this linear

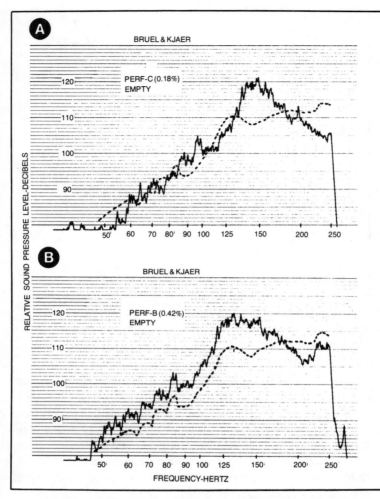

Fig. 10-28. One third octave pink noise sweeps showing sound pressure inside the Helmholtz resonator (solid recorder trace) and sound pressure outside at the resonator face (broken line). (A) 0.18%, (B) 0.42%, and (C) 1.05%, no glass fiber in the box, (D) slat type resonator without glass fiber.

pressure form we have a better comparison between the various conditions under examination. For example, Fig. 10-29A clearly shows the progression in resonance frequency with perforation percentage.

The curves of Fig. 10-29B show the effect of 4″ of 723 Fiberglas at the rear of the box and Fig. 10-29C shows the effect of the same Fiberglas pressed against the inside face of the perforated cover.

204

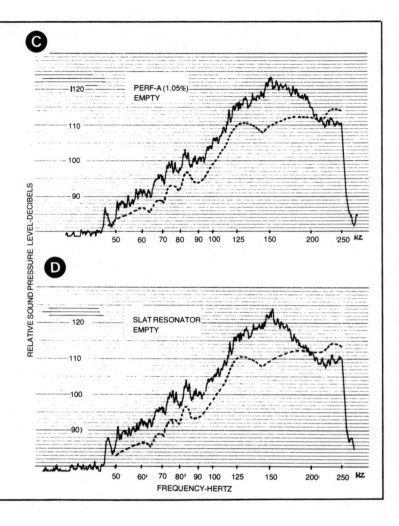

The vertical scales of Fig. 10-29A, B, and C are directly comparable, inviting comparison. The presence of glass fiber at the rear of the box, Fig. 10-29B, reduces peak pressure about 30% from the empty condition. Against the inside surface of the cover the reduction is nearer 60%. The significance of this will be discussed later.

The effect of perforation percentage in changing the peak frequency is well pronounced in Figs. 10-29A and B. When the

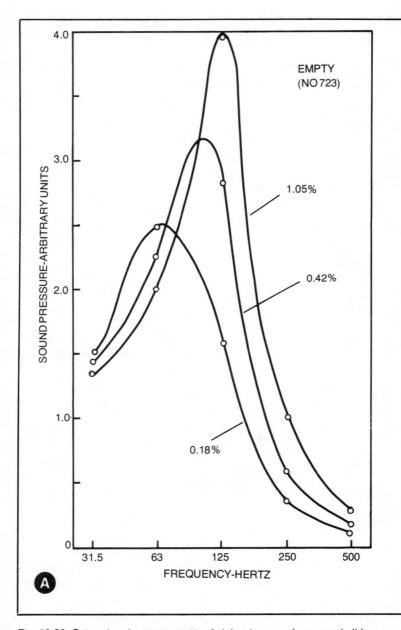

Fig. 10-29. Octave band measurements of pink noise sound pressure buildup inside of perforated face type of Helmholtz resonator near resonance: (A) no glass fiber inside, (B) 4″ 723 Fiberglas against rear of box, and (C) 4″ of Fiberglas placed against inside face of perforated cover. Sound pressure is relative to pressure at the outside face of the resonator.

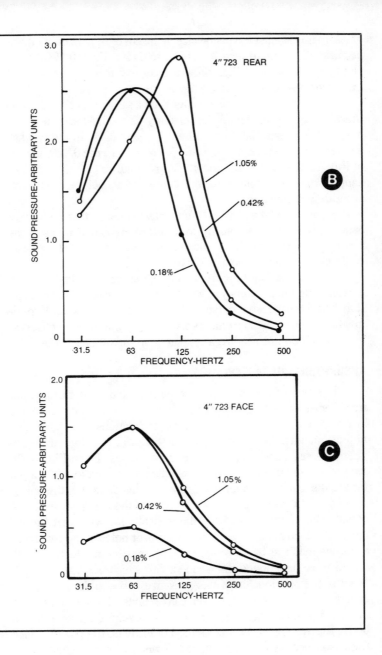

207

glass fiber is against the inside face of the cover, however, all three perforations seem to resonate closer to 63 Hz, Fig. 10-29C. In the slat type resonator of Fig. 10-30 the curve for glass fiber against the cover also peaks near 63 Hz.

In octave band measurements such as Figs. 10-29 and 10-30 there is often considerable uncertainty as to the location of the peak. This is especially true when the peak falls "in the cracks" between two octaves. Good examples of this are the 0.42% curves of Fig. 10-29A and B and the 4" 723 rear curve of Fig. 10-30. If, happily, the peak falls near the center of an octave it is more accurately delineated.

Figures 10-29 and 10-30 demonstrate clearly that the octave band approach can reasonably well determine experimentally the resonance frequency. If a ⅓ octave filter set were used the peak could be found even more accurately.

The octave band approach thus gives an excellent tool for determining the frequency at which actual resonance peaks occur. It avoids the excessive detail of the swept sine approach. Experimentally, octave measurements are simple point by point observations which do not require a graphic level or other recorder.

ABSORPTION CHARACTERISTICS

The methods we have just explored help us to determine the frequency to which a perforated or slat type of Helmholtz resonator is tuned with greater accuracy than calculations. We have also noted that introducing an absorbent material into the air cavity shifts the resonance peak to lower frequencies. From Fig. 10-30 we can also deduce that a given amount of absorbent material has a far greater effect when placed against the inside face of the perforated or slat cover than against the back of the box away from the cover. This is understandable theoretically knowing that the air particle velocity is greatest near the slots or holes of the cover and a fibrous material there would have its maximum effect.

So, we have found the frequency to which our box is tuned. What is the relationship of this pressure curve to the absorption curve we must have for our room calculations? To say the least, it is a relationship requiring mathematics beyond the scope of this book. But we do need some method of bridging the gap, for once the resonance frequency is pinned down, absorption coefficients are needed for room design.

Figure 10-31 presents the absorption coefficients reported by Mankovsky[69] for the resonator constructions listed in Table 10-3.

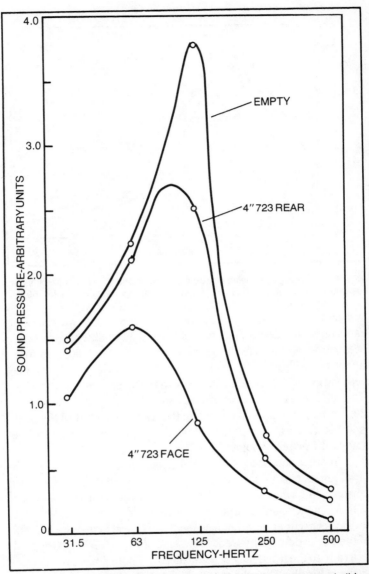

Fig. 10-30. Octave band measurements of pink noise sound pressure buildup within a slat type of Helmholtz resonator near resonance with no glass fiber inside, and with 4″ of 723 Fiberglas against rear of box and against inside surface of the slats.

The depth of his box (200 mm) is the same as ours (7-⅞″) and, except for Perf-A, the perforation percentages are close. Each of the three curves of Fig. 10-31 follows the same laws of resonance

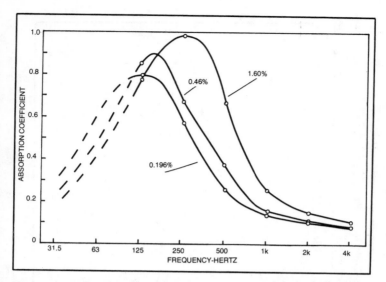

Fig. 10-31. Absorption coefficients of three perforated face type of Helmholtz resonators described in Table 10-3 and reported by Mankovsky (Ref. 69).

and their shapes are basically similar. As a result we can shift experimentally determined data of Fig. 10-31 to other frequencies without doing violence to the laws of nature. The 1.60% perforation curve of Fig. 10-31 peaks at 250 Hz. What if we wanted one peaking at 500 Hz? The easiest way to find coefficients for the 500 Hz peak is to lay a sheet of tracing paper over Fig. 10-31 and trace the 1.60% perf curve on it. Sliding the tracing paper until the peak is at 500 Hz enables one to read off absorption coefficients for the standard frequency points.

Another approach is through the normalized curve. The data of all three curves of Fig. 10-31 are replotted in Fig. 10-32 in normalized form. This requires a simple arithmetic operation on both frequency and absorption coefficient. Note in Fig. 10-31 that the peaks occur at different frequencies. Note also that the heights of the peaks differ. The frequency scale of Fig. 10-32 is expressed as a ratio to the frequency of resonance, f_o. For example, the 0.46% curve of Fig. 10-31 peaks at 150 Hz. The 500 Hz point of this curve is plotted on the normalized coordinates of Fig. 10-32 as f/f_o or $500/150 = 3.33$ on the horizontal axis. Because the curve peaks at 0.9 it needs a vertical adjustment as well. The 500 Hz absorption coefficient is 0.39. Dividing this by the 0.9 peak factor gives 0.43 as the normalized coefficient for 500 Hz. This point is then plotted as the cross at normalized frequency 3.33 and normalized coefficient

210

of 0.43. Carrying this process through point by point for the 1.60%, 0.46%, and the 0.196% cases gives us the "standard" shaped absorption curve of Fig. 10-32. It can be called a sort of universal resonance curve applicable to 200 mm (7-⅞") depth and 100 mm of absorption material inside. Mankovsky does not tell us whether the filler is against the face or in the rear of the box. It would appear from our pressure data that it was in the rear of the box.

CALCULATION OF RESONANCE FREQUENCY

We need both prediction and verification in designing Helmholtz resonators for treatment of studios and listening rooms. Prediction of resonance frequency is accomplished by calculation, but the accuracy isn't all we would hope for. By glancing at Table 10-4 we can also conclude that there are numerous possible slips betwixt computed and experimentally determined resonance peaks, although we see good agreement between the octave measurements on the empty resonators and the calculated resonance frequencies. With this realistic outlook we can do our computing without taking the precise looking figure on the calculator as

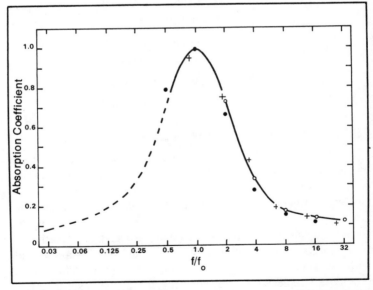

Fig. 10-32. Normalized absorption curve developed from the three Mankovsky measurements on perforated face absorbers having a depth of 200mm and with 100mm of absorbent: open circles 1.60%, black circles 0.46%, crosses 0.196% perforation. The broken line has been drawn on a basis of strict symmetry which does not always hold for resonance phenomena.

absolute truth. Michael Rettinger's book and papers[72-75] are excellent guides in such calculations.

For perforated type Helmholtz absorbers the frequency of resonance is given by:

$$f_o = 200 \sqrt{\frac{p}{(D)\,(t)}} \qquad (10\text{-}3)$$

where, f_o = frequency of resonance, Hz

        p = perforation percentage

        t = effective hole length, inches

          = (panel thickness) + 0.8 (hole diameter), approx.

      D = air space depth, inches.

For slat type Helmholtz absorbers the resonance frequency is given by:

$$f_o = 2160 \sqrt{\frac{r}{(d)\,(D) + (w + r)}} \qquad (10\text{-}4)$$

where, f_o = frequency of resonance, Hz

            r = slot width, inches

           w = width of slat, inches

            D = airspace depth, inches

            d = effective depth of slot, inches

              = 1.2 (thickness of slat), approx.

VERIFICATION OF RESONANCE FREQUENCY

As noted in Table 10-4 the resonance frequencies determined by the various methods (octave, ⅓ octave, sine sweep) do not agree in many cases, nor do they always agree with the calculated values. On top of that, whether the absorbent layer is against the face or the back or absent altogether also affects the result. Greatest confidence should be placed in the octave measurements as they are free of problems of calibration and stability associated with the swept sine and swept ⅓ octave methods. Octave tests provide a simple method of verifying the actual resonance frequency of the empty box. Once the frequency is determined the next problem is estimation of absorption coefficients.

The absorption measurements reported by Mankovsky were on boxes containing 100mm (4 inches) of "PP80" absorbent. He does not define PP80, mention its density, nor its location in the box. Helmholtz resonators currently designed for studios usually specify only one inch, or at the most two inches, of absorbent, often of 3 lb/cu ft density such as Owens-Corning Type 723 Fiberglas or Johns-Manville 1000 Series Spin-Glas (both proprietary names).

MAXIMUM PERFORMANCE

For maximum random incidence absorption Callaway and Ramer[76] have pointed out that there is a distinct advantage in spacing the absorbent material a short distance from the inside surface of the perforated facing. The amount of this spacing varies with percent perforation and density of absorbent. For a typical absorber having about 1% perforation percentage and using a glass fiber of 3 lb/cu ft density, the glass fiber should be spaced ¼″ from the perforated cover for maximum absorption.

RESONATOR Q

The sharpness of resonance curves is often described in terms of Q, or dissipation factor. This Q of a resonance system can be defined as:

$$Q = \frac{f_0}{\Delta f} \qquad (10\text{-}5)$$

where f_0 is the frequency at resonance and Δf is the width of the resonance curve at the half power (-3 dB) points. The Coca Cola

Table 10-4. Comparison of Resonance Frequencies.

	Percent Perforation						Slat
	Perf-A		Perf-B		Perf-C		
Mankovsky Ours	1.60%	1.05%	0.46%	0.42%	0.19%	0.18%	
Mankovsky Reverb. chamber msts.	250 Hz		150 Hz		125 Hz		
Ours, Octave Empty		125 Hz		100 Hz		65 Hz	125 Hz
4″Filler at face		63 Hz		63 Hz		63 Hz	63 Hz
4″ Filler at rear		110 Hz		72 Hz		58 Hz	95 Hz
Ours, 1/3 Octave Empty		150 Hz		130 Hz		140 Hz	150 Hz
Ours, Sine Empty		130 Hz		90 Hz		70 Hz	170 Hz
Calculated	169 Hz	126 Hz	91 Hz	80 Hz	59 Hz	52 Hz	107 Hz

bottles referred to earlier resonated at 185 Hz and had a bandwidth of 0.67 Hz resulting in a truly phenomenal Q of 185/0.67 or 276. Qs of this magnitude in a studio could result in "ringing" problems. The Qs of the three experimental perforations and the single slat facing vary from 1 to 5 through all conditions of internal absorbent including empty, quite different from Coke bottles.

PASSIVE VS. ACTIVE ABSORBERS

An opinion has been expressed that for critical listening rooms the treatment should be only with dissipative absorbers such as glass fiber and not with reactive type such as panel and Helmholtz vibrating systems. A consultant in Hertfordshire, England, claims that reactive absorbers introduce sound into the room which is not directly related to the original sound or its normal pattern of reflections.[77] This is the only negative note on resonance type absorbers the writer has encountered in the literature in the face of innumerable successful applications of Helmholtz resonators to studios and control rooms around the world. Perhaps the last word on this subject is yet to be heard.

REVERBERATION CONTROL

The predominant use of Helmholtz type units is in providing tunable absorption to correct for deficiencies in other types of absorbers. A uniform reverberation time throughout the audible spectrum is generally taken as the optimum. With such a reverberation characteristic all components of speech and music die away at the same rate. The studio, control room, and listening room design problem is centered in the fact that commonly available materials do not have uniform absorption throughout the band. This requires careful apportioning of many types of absorbers to finally arrive at the flat reverberation characteristic. Helmholtz units are invaluable in supplying needed absorption at low frequencies to make up for deficiencies of carpets, drapes, acoustical tile and the like. They are also useful in filling in sagging midband absorption. The incorporation of Helmholtz resonators in specific design problems has been treated in an elementary form in great detail elsewhere.[109]

With all equipment in place for the other measurements for this chapter, the temptation to answer the question, "Would a single perforated absorber affect the reverberation time of this small measuring room noticeably?", proved to be irresistable. Five decay measurements with octave bands of noise were made at each frequency both with empty Perf-A resonator in the room and not in

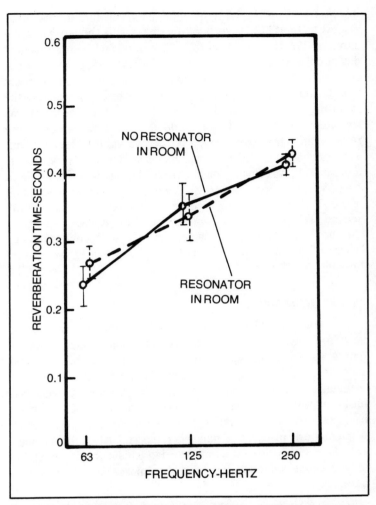

Fig. 10-33. The effect of a single 4.8 sq. ft. Helmholtz resonator on the reverberation time of a small room. The vertical lines indicate plus and minus one standard deviation from the mean value.

the room, everything else being held constant. The results are shown in Fig. 10-33. The effect at 125 Hz is more or less significant (we shall leave to the statistician just how significant) but what do you expect of only 4.8 sq ft of absorber tuned to this frequency?

RINGING OF EMPTY RESONATOR

The highest Q encountered in these experimental units is of the order of Q = 5. The question comes to mind, "Is it possible that it 'rings'?". With the microphone inside in the usual place(Fig. 10-18) the reverberation time inside the box was measured at 125 Hz and found to be 0.201 second with a standard deviation of 0.027 with 10 decays. This was done in the most convenient way with the resonator in the small measuring room next to the equipment room. To quiet the haunting fear that the room may have affected the measurements, the operation was repeated with the resonator outdoors. The mean value of the reverberation time over another ten decays was 0.222 second with a standard deviation of 0.023 which indicated that the room effect, if there at all, was small. Combining all 20 measurements yielded a mean reverberation time of 0.212 second with a standard deviation of 0.027. Compared to the room reverberation time of about 0.35 second at 125 Hz it would appear that the box decay was truly measured and found reasonable. Any ringing problem of our "ringiest" resonator which would show up in long reverberation time did not materialize.

TAMING ROOM MODES

Another use of Helmholtz resonators is to apply sharply tuned units to control particularly troublesome room modes. This requires a sophisticated approach, both in identifying the mode and then in controlling it.[78] The volume of the resonator must be mathematically determined to fit the job. Of course it would be futile to place the resonator at a node where that particular mode has a zero effect, so even the placement becomes important, and when properly placed, fine tuning is required. After all this, suitable damping(absorbent) is introduced to achieve the proper Q.But it can be done and resonators have been successfully used for this purpose.

INCREASING REVERBERATION TIME

Low Q Helmholtz resonators are capable of shortening reverberation time by increasing absorption. High Q resonators can increase reverberation time through storage of energy as described by Gilford.[79] To achieve the high Qs necessary, plywood, particle board, masonite and other such materials must be abandoned and ceramics, plaster, concrete, etc., used in resonator construction. By proper tuning of the resonators the increase in reverberation time can be placed where needed in regard to frequency.

Diffusion of Sound 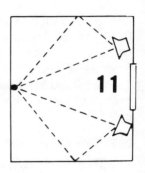 11

There are many ways to insult the owner of a recording studio, but one of the surest is to tell him that the diffusion of the sound in his studio is inadequate. Neither the insultor nor the insultee may know much about diffusion of sound except that it is a much sought after quality made even more desirable by the aura of mystery surrounding it. There is really only one sure fire method of achieving good diffusion and that is to make your room very large. Large spaces have closely spaced modes. The problem with small spaces such as the average recording studio, control room, or music listening room is that modal spacings below 300 Hz guarantee a sound field far from diffuse. "Hot spots" and "holes" of sound intensity are, in one degree or another, inevitable, but minimizing them will surely make it easier to place microphones and to realize satisfactory listening conditions.

THE PERFECTLY DIFFUSE SOUND FIELD

Even though unattainable, it is instructive to consider the characteristics of a perfectly diffuse sound field. Randall and Ward[80] have given us a list of these:

■ The frequency and spatial irregularities obtained from steady state measurements must be negligible.

■ Beats in the decay characteristic must be negligible.

■ Decays must be perfectly exponential, i.e., they must be straight lines on a logarithmic scale.

■ Reverberation time will be the same at all positions in the room.

■ The character of the decay will be essentially the same for different frequencies.

■ The character of the decay will be independent of directional characteristics of the measuring microphone.

These six factors are observationally oriented. A good physicist specializing in acoustics would stress fundamental and basic factors in his definition of a diffuse sound field such as energy density, energy flow, superposition of an infinite number of plane progressive waves, and so on. The six characteristics suggested by Randall and Ward point us to practical ways of obtaining solid evidence for judging just how diffuse the sound field of a given room might be.

DIFFUSION EVALUATION

There is nothing quite as upsetting as viewing one's first attempt at measuring the "frequency response" of a room. To obtain the frequency response of an amplifier a variable frequency signal is put in the front end and the output is observed to see how flat the response is. The same general approach can be applied to a room by injecting the variable frequency signal into "the front end" by means of a loudspeaker and noting the "output" picked up by a microphone located elsewhere in the reverberant field of the room.

Steady State Measurements

Figure 11-1 is a graphic level recorder tracing of the steady state response of a studio having a volume of 12,000 cu. ft. In this case the loudspeaker was in one lower corner of the room and the microphone was at the upper diagonal corner about one foot from each of the three surfaces. These positions were chosen on the basis that all room modes terminate in the corners and that we want all modes represented in the trace. The fluctuations in this response cover a range of about 35 dB over the linear 30 to 250 Hz sweep. The nulls are very narrow and the narrower peaks show evidence of being single modes as the mode bandwidth of this room is close to 4 Hz. The wider peaks are the combined effect of several adjacent modes. The rise from 30 to 50 Hz is due primarily to loudspeaker response and should not be charged against the room. The rest is primarily room effect.

The response of Fig. 11-1 is typical of even the best studios. Such variations in response are, of course, evidence of a sound field

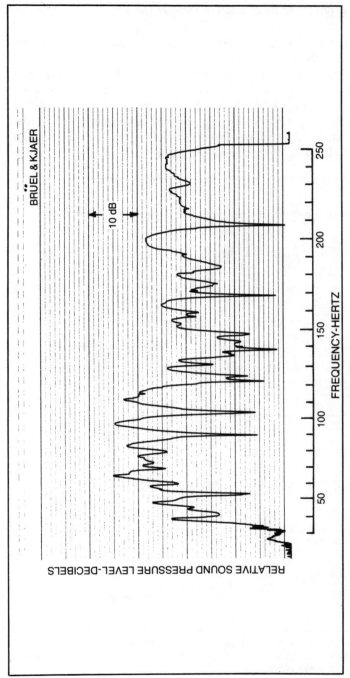

Fig. 11-1. Slowly swept sine wave sound transmission response of a 12,000 cu. ft. studio. Fluctuations of this magnitude which characterize the best of studios are evidence of non-diffuse conditions.

219

that is not perfectly diffused. A steady state response such as this taken in an anechoic room would still show variations, but of lower amplitude. A very live room, such as a reverberation chamber, would show even greater variations.

Figure 11-1 illustrates one way to obtain the steady state response of a room. Another is to traverse the microphone while holding the loudspeaker frequency constant. Both methods reveal the same deviations from a truly homogenous sound field. Thus we see that Randall and Ward's criterion of negligible frequency and spatial irregularities does not qualify the studio of Fig. 11-1 or, in

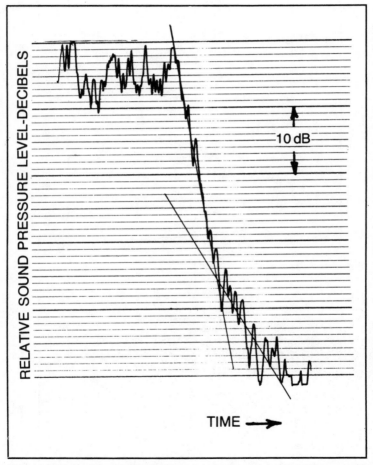

Fig. 11-2. Typical double slope decay, evidence of a lack of diffuse sound conditions. The slower decaying final slope is probably due to modes which encounter lower absorption.

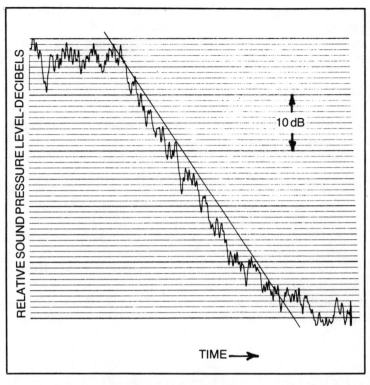

Fig. 11-3. The non-exponential form of this decay, taken in a 400 seat chapel, is attributed to acoustically coupled spaces. Lack of a diffuse sound field is indicated.

fact, any practical recording studio as having a perfectly diffuse field.

Decay Beats

By referring back to Fig. 8-9 we can compare the smoothness of the reverberation decay for the eight octaves from 63 Hz to 8 kHz. In general the smoothness of the decay increases as frequency is increased. The reason for this, as explained in Chapter 8, is that the number of modes within an octave span increases greatly with frequency and the greater the mode density, the smoother their average effect. Beats in the decay are greatest at 63 Hz and 125 Hz. The decays of Fig. 8-9 indicate that the diffusion of sound in this particular studio is about as good as can be achieved practically. It is the beat information on the low-frequency reverberation decay that makes possible a judgement on the degree of diffusion

221

prevailing. Reverberation time measuring devices which yield information only on the slope and not the shape of the decay pass over information which most consultants consider important in evaluating the diffuseness of a space.

Exponential Decay

A truly exponential decay is a straight line on a level vs. time plot and the slope of the line may be described either as a decay rate in decibels per second or as reverberation time in seconds. The decay of the 250 Hz octavé band of noise pictured in Fig. 11-2 has two exponential slopes, the initial slope gives a reverberation time of 0.35 second and the final slope a reverberation time of 1.22 seconds. The slow decay that finally takes over once the level is low enough is probably a specific mode or group of modes encountering low absorption either by striking the absorbent at grazing angles or striking where there is little absorption. This is typical of one type of non-exponential decay or, stated more precisely, of a dual exponential decay.

Another type of non-exponential decay is illustrated in Fig. 11-3. The deviations from the straight line connecting the beginning and end of the decay are considerable. This is a decay of an octave band of noise centered on 250 Hz in a 400-seat chapel, poorly isolated from an adjoining room. Decays taken in the presence of acoustically coupled spaces are characteristically concave upward, such as Fig. 11-3, and often the deviations from the straight line are even greater. When the decay traces are non-exponential, i.e., depart from a straight line in a level vs. time plot, we must conclude that true diffuse conditions do not prevail.

Spatial Uniformity of RT60

When reverberation time for a given frequency is reported it is usually the average of multiple observations of each of several positions in the room. This is the pragmatic way of admitting that reverberatory conditions differ from place to place in the room. Figure 11-4 shows the results of actual measurements in a small (22,000 cu.ft.) video studio. The multiple uses of the space required variable reverberation time which was accomplished by hinged wall panels which can be closed, revealing absorbent sides, or opened, revealing reflecting sides. Multiple reverberation decays were recorded at the same three microphone positions for both "panels open" and "panels closed" conditions. The circles are the average values and the light lines represent average reverbera-

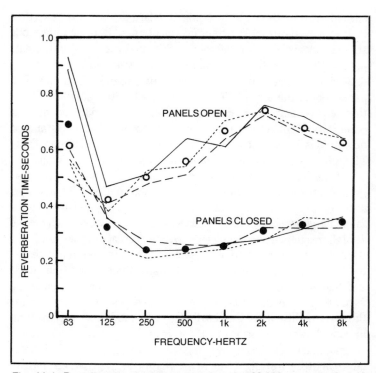

Fig. 11-4. Reverberation time characteristics of a 22,000 cu. ft. studio with acoustics adjustable by hinged panels, absorbent on one side and reflective on the other. At each frequency the variation of average reverberation time at each of the three positions, especially at low frequencies, indicates non-diffuse conditions.

tion time at each of the three positions. It is evident that there is considerable variation which means that the sound field of the room is not completely homogenous during this transient decay period. Inhomogeneities of the sound field are one reason for RT60 values varying from point to point in the room, but there are other factors as well. Uncertainties in fitting a straight line to the decay also contribute to the spread of the data, but this effect should be relatively constant from one position to another. It seems reasonable to conclude that spatial variations of RT60 are related to the degree of diffusion in the space.

Standard deviations of the reverberation times give us a measure of the spread of the data as measured at different positions in a room. When we calculate an average value, all evidence of the spread of the data going into the average is lost. The standard deviation is the statistician's way of keeping an eye on the data spread. The method of calculating standard deviation is described in the manuals of most scientific calculators. Plus or minus one standard deviation from the mean value embraces 63% of the data points if the distribution is normal (Gaussian) and reverberation data should qualify reasonably well. In Table 11-1 for 500 Hz, panels open, the mean RT60 is 0.56 second with a standard deviation of 0.06 second. For a normal distribution, 63% of the data points would fall between 0.50 and 0.62 second. That 0.06 standard deviation is 11% of the 0.56 mean. The percentages listed in Table 11-1 give us a rough appraisal of the precision of the mean.

In order to view the columns of percentages in Table 11-1 graphically they are plotted in Fig. 11-5. Variability of RT60 values at the higher frequencies settle down to reasonably constant values in the neighborhood of 3 to 6%. Because we know that each octave at high frequencies contains an extremely great number of modes which results in smooth decays, we can conclude confidently that at the higher audible frequencies essentially diffuse conditions exist and that the 3-6% variability is normal experimental measuring variations. At the low frequencies, however, the high percentages (high variabilities) are the result of greater mode spacing resulting in considerable variations of RT60 from one position to another. We must also admit that these high percentages include the uncertainty in fitting a straight line to the wiggly decays characteristic of low frequencies. However, a glance back at Fig. 11-4 shows that great differences in RT60 are noted between the three measuring positions. For this 22,000 cu.ft. studio for two different conditions of absorbence (panels open/closed), there is poor diffusion at 63 Hz, somewhat better at 125 Hz, and reasonably good at 250 Hz and above.

Decay Shapes

If all decays have the same character at all frequencies and that character is a smooth decay, complete diffusion prevails. In our real world the decays of Fig. 8-9 are more common with significant changes in character, especially for the 63 Hz and 125 Hz decays.

Microphone Directivity

A method of appraising room diffusion is to rotate a highly directional microphone in various planes, recording its output to the constant excitation of the room. This has been applied, with some success, to large spaces but the method is ill adapted to smaller recording studios, control rooms, and listening rooms. In

Table 11-1. Reverberation Time of Small Video Studio.

Octave band center frequency Hz	PANELS OPEN			PANELS CLOSED		
	Mean	Std. Dev.	Std. Dev. % of mean	Mean	Std. Dev.	Std. Dev. % of mean
63	0.61	0.19	31.	0.69	0.18	26.
125	0.42	0.05	12.	0.32	0.06	19.
250	0.50	0.05	10.	0.24	0.02	8.
500	0.56	0.06	11.	0.24	0.01	4.
1k	0.67	0.03	5.	0.26	0.01	4.
2k	0.75	0.04	5.	0.31	0.02	7.
4k	0.68	0.03	4.	0.33	0.02	6.
8k	0.63	0.02	3.	0.34	0.02	6.

principle, however, in a totally homogenous sound field, a highly directional microphone pointed in any direction should pick up a constant signal.

ROOM SHAPE

How can we build or treat a room so that the sound field in the room will be as diffuse as possible? This opens up a field in which there are strong opinions - some of them supported by quite convincing experiments and some just strong without such support.

There are many possible shapes of rooms. Aside from the general desirability of a flat floor in this gravity stricken world, walls can be splayed, ceiling inclined, cylindrical or polygonal shapes employed. Many shapes can be eliminated because they focus sound and focussing is the opposite of diffusing. For example, parabolic shapes have beautifully sharp focal points, cylindrical concavities less sharp but nonetheless concentrations of energy. Even polygonal concave walls of 4, 5, 6, or 8 sides approach a circle and result in concentrations of sound in some areas at the expense of other areas.

The popularity of rectangular rooms is due, in part, to economy of construction, but it has its acoustical advantages. The axial, tangential, and oblique modes can be, with some effort, calculated and their distribution studied. To a first approximation, a good approach is to consider only the more dominant axial modes which is a very simple calculation. Degeneracies (mode pile-ups) can be spotted and other room faults revealed.

As we shall see, the relative proportioning of length, width, and height of a sound sensitive room is most important. If plans are being made for constructing such a room, there are usually ideas on floor space requirements, but where should one start in regard to room proportions? We know that cubical rooms are anathema. The literature is full of quasi-scientific guesses and, later, statistical analyses of room proportions to give good mode distribution. None of them come right out and say, "This is best". Bolt[81] gives a range of room proportions producing smoothest frequency response at low frequencies in small rectangular rooms. Volkmann's[82] 2:3:5 proportion, in favor forty years ago, falls within the Bolt range as does Boner's[83] $1:\sqrt[3]{2}:\sqrt[3]{4}$ ratio. Sepmeyer[84] published a computer statistical study 21 years after Bolt's paper which yields several favorable ratios. An even later paper by Louden[85] lists 125 dimension ratios arranged in descending order of room acoustical quality. Table 11-2 presents the best proportions suggested by these papers.

One cannot tell by looking at a room's dimensional ratio whether it is desirable or not and we would prefer to make the evaluation ourselves rather than just taking someone's word for it. Assuming a room height of 10 ft., the other two dimensions may readily be calculated from the ratios of Table 11-2. Once we have the room dimensions a mode analysis such as Fig. 11-6 can be made for each. This has been done and these modes are plotted in Fig. 11-7. Each is keyed into Table 11-2 for source identification. All of these are relatively small rooms and therefore suffer the same fate

Table 11-2. Rectangular Room Dimension Ratios for Favorable Mode Distribution.

Author		Height	Width	Length	In Bolt's Area?
1. Sepmeyer	A	1.00	1.14	1.39	No
(Ref 84)	B	1.00	1.28	1.54	Yes
	C	1.00	1.60	2.33	Yes
2. Louden	A	1.00	1.4	1.9	Yes
3 best					
ratios	B	1.00	1.3	1.9	No
(Ref 85)	C	1.00	1.5	2.1	Yes
3. Volkmann 2:3:5 (Ref 82)		1.00	1.5	2.5	Yes
4. Boner $1:\sqrt[3]{2}:\sqrt[3]{4}$ (Ref 83)		1.00	1.26	1.59	Yes

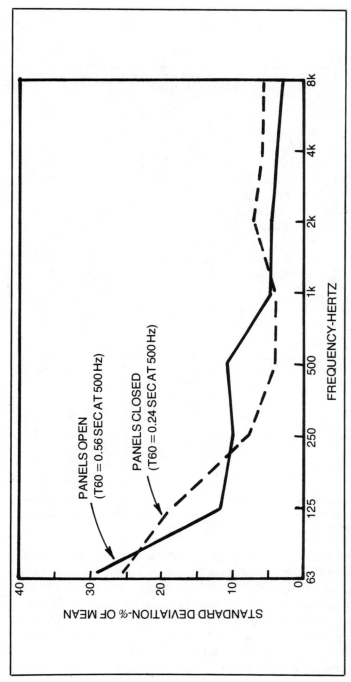

Fig. 11-5. Closer examination of the reverberation time variations of the studio of Fig. 11-4. The standard deviation, expressed as a percentage of the mean value, shows lack of diffusion, especially below 250 Hz.

227

ROOM DESCRIPTION _RECORDING STUDIO_   DATE _9 July 1980_

	LENGTH	WIDTH	HEIGHT	Arranged in ascending order	Diff.
	L=_19_ ft _5_in L=_19.417_ft f_1= 565/L	W=_14_ft _2_in W=_14.17_ft f_1= 565/W	H=_8_ft_11_in H=_8.92_ft f_1= 565/H	29.1	
					10.8
				39.9	
					18.3
f_1	29.1	39.9	63.3	58.2	
					5.1
f_2	58.2	79.7	126.7	63.3	
					16.4
f_3	87.3	119.6	190.0	79.7	
					7.6
f_4	116.4	159.5	253.4	87.3	
					29.1
f_5	145.5	199.4	316.7	116.4	
					3.2
f_6	174.6	239.2		119.6	
					7.1
f_7	203.7	279.1		126.7	
					18.8
f_8	232.8	319.0		145.5	
					14.0
f_9	261.9			159.5	
					15.1
f_{10}	291.0			174.6	
					15.4
f_{11}	320.1			190.0	
					9.4
f_{12}				199.4	
					4.3
f_{13}				203.7	
					29.1
f_{14}				232.8	
					6.4
f_{15}				239.2	
					14.2
				253.4	
					8.5
				261.9	
					17.2
				279.1	
					11.9
				291.0	
					25.7
				316.7	

Stop at 300 Hz

Fig. 11-6. A convenient data form for studying the effects of room proportions on the distribution of axial modes.

of having axial mode spacings in frequency greater than desired. The more uniform the spacing the better. Degeneracies or mode coincidences are a potential problem and they are identified by the "2" or "3" above them to indicate the number piled up. Modes very close together, even though not actually coincident, can also present problems. With these rules to follow, which of the 8 "best" distributions of Fig. 11-7 are really the best? Which the worst? First, we reject line 3 with two triple coincidences greatly spaced from neighbors. Next, line 2-C is eliminated because of three double coincidences associated with some quite wide spacings. We can neglect the effect of the double coincidences near 280 Hz in line 1-C and line 2-A because colorations are rarely experienced above 200 Hz. Aside from the two rejected outright, there is little to choose between the remainder. All have flaws. All would probably serve quite well, alerted as we are to potential problems here and there. This simple approach of studying the axial mode distribution

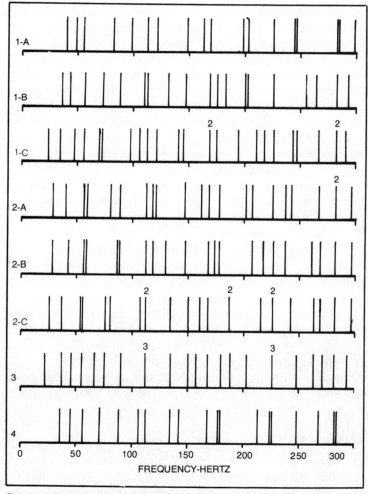

Fig. 11-7. Plots of axial mode distribution for the eight "best" room proportions of Table 11-2. The small numbers indicate the number of modes coincident at those particular frequencies. A room height of 10 ft is assumed.

has the advantage of paying attention to the dominant axial modes in the knowledge that the weaker tangential and oblique modes can only help by filling in between the more widely spaced axial modes.

Figure 11-6 illustrates a data form which makes easy the study of the axial modes of a room. The analysis of the results requires some experience and a few rules of thumb. A primary goal is to avoid coincidences or pile-ups of axial modes. For example, if a cubical space were analyzed on the form of Fig. 11-6 all three columns would be identical; the three fundamentals and all harmonics would coincide. This would result in a triple-whammy at each modal frequency and great gaps between. Unquestionably, sound in such a cubical space would be highly colored and acoustically very poor. The room of Fig. 11-6 has 22 modes between 29.1 and 316.7 Hz. If evenly spaced the spacing would be about 13 Hz but we note that spacings vary from 3.2 to 29.1 Hz. However, there are no coincidences - the closest pair are 3.2 Hz apart. If a new room is to be constructed, one has the freedom on paper of moving a wall this way or that or raising or lowering the ceiling a bit to improve distribution. The particular room proportions of Fig. 11-6 represent the end product of hours of cut and try. While not representing it as the best proportioning possible, it is quite certain that this room, properly treated, will yield good, uncolored sound. The proper starting point is proper room proportions and it is downhill from there on.

In adapting an existing space the freedom of shifting walls on paper is not present. A study of the axial modes as per Fig. 11-6, however, can still be very helpful. For example, if such a study reveals problems, and space permits, a new wall might improve the modal situation markedly. By splaying this wall, other advantages to be discussed later may accrue. If the study points to a coincidence at 158 Hz, well separated from neighbors, one is alerted to potential future problems with an understanding of the cause. There is always the possibility of introducing a Helmholtz resonator tuned to the offending coincidence to control its effect. All these things are related to our subject of sound diffusion.

SPLAYING WALLS

Splaying of one or two walls of a sound sensitive room does not eliminate modal problems, although it may shift them slightly and tend somewhat toward better diffusion.[86] In new construction splayed walls are inexpensive but may be quite expensive in adapting an existing space. Wall splaying is one way of improving general room diffusion, although its effect is nominal. Flutter echoes can definitely be controlled by canting one of two opposing walls. The amount of splaying is usually between 1 foot in 20 feet and 1 foot in 10 feet.

GEOMETRICAL IRREGULARITIES

Many studies have been made on what type of wall protuberances provide the best diffusing effect. Somerville and Ward[87] reported years ago that geometrical diffusing elements reduced fluctuations in a swept sine steady state transmission test. The depth of such geometrical diffusers must be at least 1/7th of a wavelength before their effect is felt. They studied cylindrical, triangular, and rectangular elements and found that the straight sides of the rectangular shaped diffuser provided the greatest effect for both steady state and transient phenomena. BBC experience indicates superior subjective acoustical properties in studios and concert halls in which rectangular ornamentation in the form of coffering is used extensively.

ABSORBENT IN PATCHES

Applying all the absorbent in a room on one or two surfaces does not result in a diffuse condition, nor is the absorbent used most effectively. Let us consider the results of an experiment which shows the effect of distributing the absorbent.[80] The experimental room is approximately a 10 ft. cube and it was tiled (not an ideal recording or listening room, but acceptable for this experiment). Test 1: reverberation time for the bare room was measured and found to be 1.65 seconds at 2 kHz. Test 2: a common commercial absorber was applied to 65% of one wall (65 sq. ft.) and the reverberation time at the same frequency was found to be about 1.02 seconds. Test 3: the same area of absorber was divided into four sections, one piece mounted on each of four of the room's six surfaces. This brought the reverberation time down to about 0.55 second.

The startling revelation here is that the area of absorber was identical between Tests 2 and 3, the only difference was that in 3 it was in four pieces, one on each of 3 walls and one piece on the floor. By the simple expedient of dividing the absorbent and distributing it the reverberation time was cut almost in half. Cranking the values of RT60 of 1.02 and 0.55 seconds and the volume and area of the room into the Sabine equation we find that the average absorption coefficient of the room increased from 0.08 to 0.15 and the

Fig. 11-8. The use of distributed sound absorbing modules is an economical way to achieve maximum absorption as well as to contribute positively to sound diffusion in the room (courtesy of World Vision International).

number of absorption units from 48 to 89 sabins. Where did all this extra absorption come from? Testing laboratory personnel in reverberation chamber measurement of absorption coefficients have agonized over the problem for years. Their conclusion is that there is an edge effect related to diffraction of sound which makes a given sample appear to be much larger acoustically. Stated another way, the sound absorbing efficiency of 65 sq.ft. of absorbing material is only about half that of four 16 sq.ft. pieces distributed about the room and we note that the edges of the four pieces total about twice that of the single 65 sq.ft. piece. So . . . one advantage of distributing the absorbent in a room is that its sound absorbing efficiency is greatly increased, at least at certain frequencies. Warning: the above statements are true for 2 kHz, but at 700 Hz and 8 kHz the difference between one large piece and four distributed pieces is small.

Another significant result of distributing the absorbent is that it contributes to diffusion of sound. Patches of absorbent with reflective walls showing between the patches have the effect of altering wavefronts which improves diffusion. Sound absorbing modules in a recording studio such as in Fig. 11-8 distribute the absorbing material and, simultaneously, contribute to the diffusion of sound.

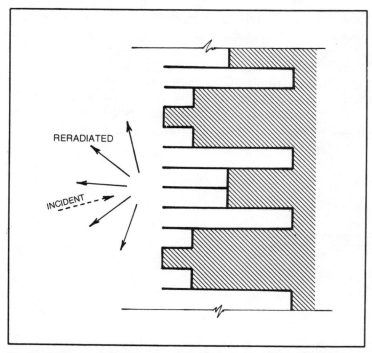

Fig. 11-9. Cross section of a possible realization of a wall of Schroeder diffusers. The depths are determined by the quarter wavelength of the design frequency and the arrangement by "Quadratic Residue" mathematical sequence.

THE SCHROEDER DIFFUSER

In a remarkable outpouring of fresh, new ideas during the past few years, M.R. Schroeder of the University of Göttingen, Germany, has opened new vistas on diffusion of sound. It is a fair guess that during the next decade his methods will be widely applied but at this writing only one application is known, the ceiling of the Town House in Wellington, New Zealand. The first proposals of his idea[88] were based on pseudo-random sequences ("Maximum-length" sequences) but more recently interest has shifted to what he calls "Quadratic Residue" sequences because of superior diffusing qualities.[89] Schroeder has applied his diffusers to music hall problems to achieve desirable lateral reflections from the ceiling.[90]

Figure 11-9 illustrates a section of what might possibly be a portion of the wall of the recording studio of the future. The sound impinges on a wall made up of cavities whose depths are determined by the quarter wavelength of the sound to be acted upon and the arrangement in accordance with the mathematical sequence. In practice, good diffusion is obtained a half octave above and a half octave below the design frequency and numerous design frequencies assure covering the desired band. Two dimensional arrays of cavities could cover a surface or the array could be rotated about a point to achieve acoustical or esthetic goals.

Quadratic residue (QR) diffusing mechanisms may or may not find extensive application in our sound recording or playback environments, but, because of the great desirability of more diffuse sound fields, the prospects are good enough to encourage close following of new developments.

Quiet Air for the Studio 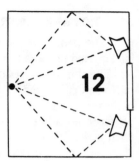 12

The background noise levels in recording studios, control rooms, and listening rooms must be kept under control if these rooms are to be of maximum use in their intended way. Hums, buzzes, rumbles, aircraft noises, auto horns tooting, dogs barking, or typewriter sounds are most incongruous if audible during a lull in the program. Such sounds might not be noticed outside the studio when they are a natural part of the situation, but during a pause of a quiet musical or speech passage they stand out like the proverbial sore pollex.

In a studio interfering sounds can come from control room monitors operating at high level or from equipment in adjacent areas. Control rooms have their own noise problems, some intruding from the outside, some generated by recorders, reverberation plates, equipment cooling fans, etc. There is one source of noise, however, that is common to all sound sensitive rooms and that is the noise coming from the air conditioning diffusers or grilles, the subject of this chapter. A certain feeling of helplessness in approaching air conditioning noise problems is widespread and quite understandable. Let's try to dispel some of the mystery surrounding the subject.

The control of air conditioning noise can be expensive. A noise specification in an air conditioning contract for a new structure can escalate the price. Alterations of an existing air conditioning system to correct high noise levels can also be expensive. It is highly desirable for studio designers to have a basic understanding of potential noise problems in air conditioning systems so that adequate control can be exercised during planning stages and supervision during installation. This applies equally to the most ambitious studio and to the budget job.

SELECTION OF NOISE CRITERION

The single most important decision having to do with background noise is the selection of the noise level goal to shoot for. The almost universal approach to this is embodied in the family of Noise Criteria (NC) contours[91] of Fig. 12-1. The selection of one of these contours establishes the goal of maximum allowable sound pressure level in each octave band. Putting the noise goal in this form makes it easily checkable by instruments. The downward slope of these contours reflects both the lower sensitivity of the human ear at low frequencies and the fact that most noises having distributed energy drop off in a similar way with frequency. To determine whether the noise in a given room meets the contour goal selected, sound pressure level readings are made in each octave and plotted on the graph of Fig. 12-1. The black dots represent such a set of measurements made with a sound level meter equipped with octave filters. The level at 1 kHz determines that a convenient single number NC-20 rating applies to this particular noise. If the NC-20 contour had been specified as the highest permissible sound pressure level in an air conditioning contract, the above installation would just barely be acceptable.

Which contour should be selected as a limit of allowable background noise in a recording studio? This depends on the general studio quality level to be maintained, on the use of the studio, and other factors. There is little point in demanding NC-15 from the air conditioning system when intrusion of traffic and other noise is higher than this. In general, NC-20 should be the highest contour that should be considered for a recording studio or listening room and NC-15 is suggested as a practical and attainable design goal for the average studio. NC-10 would be excellent and it would probably take special effort and expense to reduce all noise to this level. If one feels the urge to calibrate the senses as to just how an NC-15 or NC-20 noise really sounds, the following is suggested. Beg, borrow, buy, or rent a sound level meter with built-in octave filters. Measure the sound levels in several studios which you suspect have high noise or studios you consider quite acceptable. By the time you have measured four or five such rooms

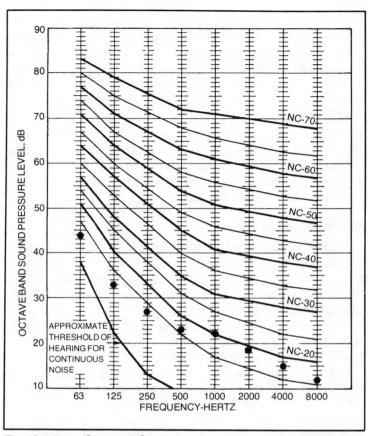

Fig. 12-1. Noise Criterion (NC) curves to be used in specifying the maximum permissible noise sound pressure level in each octave band. Measurements shown by the black spots would indicate that NC-20 contour would apply in rating that sound (after Beranek, Reference 91).

with A/C turned on and off, you will have little NC numbers in your head for ready recall in future discussions and you, too, will have become an expert.

ASHRAE

Perhaps it sounds like an Egyptian goddess but our new word for today is ASHRAE (as in ashtray). It stands for the American Society of Heating, Refrigerating, and Air-Conditioning Engineers. Although its primary purpose is to enlighten and standardize its own engineers, this highly respected organization is a prolific source of helps for the studio designer who is a novice in

A/C. In this brief chapter it is impractical to go deeply into specific design techniques, but *ASHRAE Handbook and Product Directory-1976 - Systems*[92] has its entire Chapter 35 devoted to step by step procedures in sound and vibration control. For those a bit shaky in acoustical fundamentals (e.g., how do you convert sound power to sound pressure?) a companion volume, *ASHRAE Handbook and Product Directory - 1977 - FUNDAMENTALS*,[93] is a must. For those with some engineering training and a modicum of determination, these two volumes can provide a background for intelligent dealing with the highest caliber of air conditioning contractors. For the lower caliber type this background is indispensible for avoiding big and expensive mistakes.

MACHINERY NOISE

The first step in the reduction of A/C noise in the studio is wisely locating the A/C machinery. If this is left to chance, Murphy's Law will result in the equipment room being adjacent to or on the roof directly above the studio. The wall or roof panel, vibrating like a giant diaphragm, is remarkably efficient in radiating airborne noise into the studio. So, step number one is to locate the equipment as far removed from the sound sensitive areas as possible, the next county if it can be arranged.

The next step would be the consideration of some form of isolation against structure-borne vibration. If the equipment is to rest on a concrete slab shared with the studio, and plans are being drawn up, the machinery room slab should be isolated from the main floor slab. Compressed and treated glass fiber strips are available which are suitable to separate slabs during pouring. Other precautions would include proper vibration isolation mounts, designed accurately or they will be useless or downright damaging to the situation. Flexible joints in pipes and ducts as they leave the machinery room may be advisable.

FAN NOISE

The fan is a chief, but by no means the only, contributor to A/C noise in the studio. The sound power output of the fan is largely fixed by the air volume and pressure required in the installation, but there are certainly variations between types of fans. Figure 12-2 shows the sound power output vs. frequency for eight different types of fans. The centrifugal fan with airfoil blades produces the lowest noise, but the octave centered on 500 Hz has an exceptionally high level due, no doubt, to a strong single frequency component. Fans with less than 15 blades tend to generate pure tones which may dominate their output. The fundamental frequency of this tone is (rpm/60) (number of blades) and harmonics should be expected. Incidentally, the graphs of Fig. 12-2 are plots of the data of Table 16 of Reference 2.

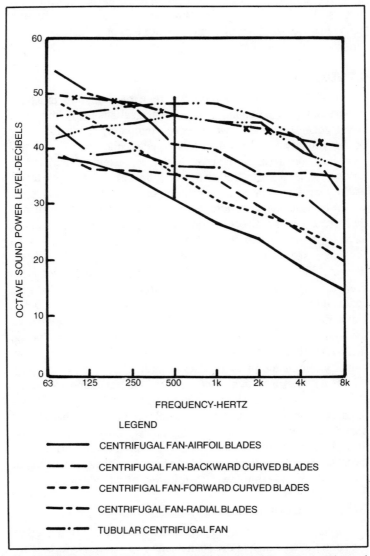

Fig. 12-2. Noise sound power output of different types of fans commonly used for air conditioning. This is a plot of the data of Table 16, Reference 92.

239

Most of the other fans of Fig. 12-2 also have single frequency tonal components but they seem to be less dominant. Fan noise, aside from the fan tones, is distributed in nature, but falling off somewhat as frequency is increased. Noise is only one point to be considered in fan selection, but perusing the manufacturer's noise specifications should receive its rightful attention along with other factors. The 20 dB spread in sound power level emphasizes this point. In general, centrifugal fans produce less noise than axial flow fans.

AIR VELOCITY

In air distribution systems the velocity of air flow is a very important factor in keeping A/C noise at a satisfactorily low level. Noise generated by air flow varies approximately as the 6th power of the velocity. As air velocity is doubled, the sound level will increase about 16 dB at the room outlet. Some authorities say that air flow noise varies as the 8th power of the velocity and give 20 dB as the figure associated with doubling or halving the air velocity.

A basic design parameter is the quantity of air the system is to deliver. There is a direct relationship between the quantity of air, air velocity, and size of duct. The velocity of the air depends upon the cross sectional area of the duct. For example, if a system delivers 500 cu ft/minute and a duct has 1 sq ft of cross sectional area, the velocity is 500 ft /min. If the area is 2 sq ft the velocity is reduced to 250 ft /min; if 0.5 sq ft velocity is increased to 1,000 ft /min. Rettinger[72] suggests an air velocity maximum of 500 ft/min for broadcast studios and this value seems about right also for top flight recording studios and other critical spaces. Specifying a low velocity eliminates many headaches later.

High pressure, high velocity, small duct systems are generally less expensive than low velocity systems. True budget systems commonly guarantee noise problems in studios because of high air velocity and its resulting high noise. Compromise may be made by flaring out the ducts just upstream from the grille. The increasing cross sectional area of the flare results in air velocity at the grille considerably lower than in the duct feeding it.

EFFECT OF TERMINAL DEVICES

Even if fan and machinery noise are sufficiently attenuated by the time the air reaches the sound sensitive room, air turbulence associated with nearby 90° bends, dampers, grilles, and diffusers can be serious noise producers as suggested by Fig. 12-3. Air flow

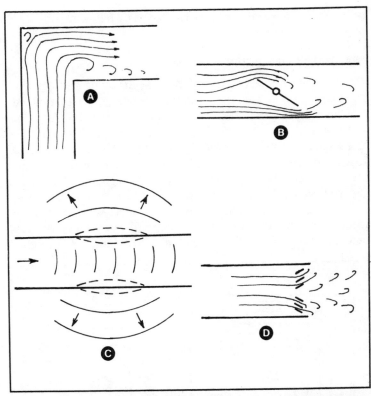

Fig. 12-3. Air turbulence caused by discontinuities in the flow path can be a serious producer of noise in (A) 90° miter bend, (B) damper used to control the quantity of air, (C) sound radiated from duct walls set into vibration by turbulence or noise inside the duct, and (D) grilles and diffusers.

in a duct can set duct walls to vibrating. Noise radiated from the duct walls can be a serious problem with exposed ducts running through a studio. Lagging the external surface with thermal type of material dampens such vibrations. Air turbulence noise can be reduced by carefully designed vanes, deflectors, and airfoils as shown in Fig. 12-4.

To emphasize the importance of such streamlining devices, here are a few numbers. Half closing an ordinary damper increases the noise level approximately 15 dB. Figure 12-5 reveals the dramatic effect, both of air velocity and shaped deflectors, in this 90° miter elbow. A 28 dB reduction of noise results from reducing air velocity from 2,000 to 700 fpm. The high frequency noise is greatly reduced at both air velocities by the use of shaped deflectors to guide the air around the right angle bend with a minimum of turbulence. These curves were calculated by following ASHRAE procedures in Reference 92.

"NATURAL" ATTENUATION

Care must be exercised to avoid expensive overdesign of an air distribution system by neglecting certain attenuation effects built into the system. There is an attenuation resulting from the fact that ducts terminate flush with wall or ceiling. The duct acts like a radio frequency transmission line by reflecting energy back toward the source when a discontinuity in impedance is encountered. Acoustical energy associated with the higher air pressure in the duct encounters the lower air pressure of the room and energy is reflected back toward the source. This effect is greatest at low frequencies and for small ducts. At the 63 Hz octave, this loss amounts to 17 dB for a 5″ duct according to Table 11 of Reference

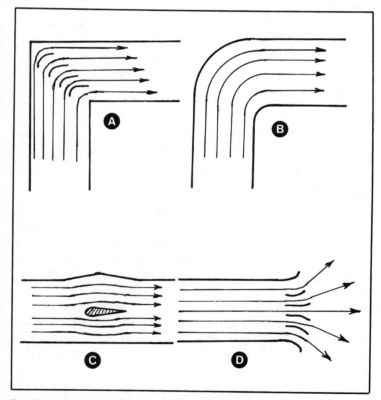

Fig. 12-4. Air turbulence noise can be materially reduced by (A) use of deflectors, (B) radius bends, (C) airfoils, and (D) carefully shaped grilles and diffusers.

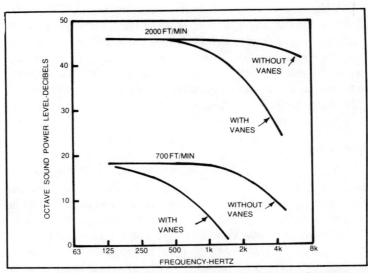

Fig. 12-5. Noise produced by a 12″ x 12″ square cross section 90° miter elbow with and without deflector vanes and at 2,000 and 700 fpm air velocities. These curves have been calculated following the procedures in Reference 92.

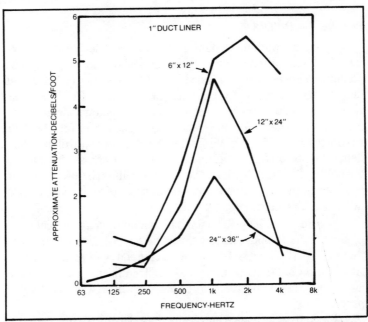

Fig. 12-6. Attenuation offered by 1″ duct liner on all four sides of a rectangular duct. Dimensions shown are the free area inside the duct, no air flow. This is a plot of data selected from Table 18, Reference 92.

92. A similar loss results at every branch or takeoff (Table 12). There is also an attenuation of noise in bare, rectangular sheet ducts due to wall flexure amounting to 0.1 or 0.2 dB per foot at low frequencies (Table 13). Round elbows introduce an attenuation, especially at the higher frequencies (Table 14). All of these losses are built into the air handling system and serve to attenuate fan and other noise coming down the duct. Its there, its free, so take it into consideration to avoid over design.

DUCT LINING

The application of sound absorbing materials to the inside surfaces of ducts is a standard method of reducing noise levels. Such lining comes in the form of rigid boards and blankets and in thicknesses of from ½″ to 2″. Such acoustical lining also serves as thermal insulation when it is required. The approximate attenuation offered by 1″ duct lining in typical rectangular ducts depends on duct size as shown in Fig. 12-6. The approximate attenuation of round ducts is given in Fig. 12-7. Duct attenuation is much lower in the round ducts than in lined rectangular ducts for comparable cross sectional areas.

PLENUM SILENCERS

A sound absorbing plenum is an economical device for achieving significant attenuation. Figure 12-8 shows a modest sized plenum chamber which, if lined with 2″ thickness of 3 lb/cu ft density glass fiber, will yield a maximum of about 21 dB attenuation. The attenuation characteristics of this plenum (calculated by Equation 3, page 35.13, Reference 92) are shown in Fig. 12-9 for two thicknesses of lining. With a lining of 4″ fiber board of the same density quite uniform absorption is obtained across the audible band. With 2″ glass fiber board attenuation falls off below 500 Hz. It is apparent that the attenuation performance of a given plenum is determined primarily upon the lining.

Figure 12-10 gives actual measurements on a practical lined plenum muffler approximately the same horizontal dimensions as that of Fig. 12-8, but only half the height and with baffles within. Attenuation of 20 dB or more was realized in this practical case above the 250 Hz octave and it solved an otherwise intolerable problem.

Plenum performance may be increased by increasing the ratio of cross sectional area of the plenum to the cross sectional area of the entrance and exit ducts, and by increasing the amount or

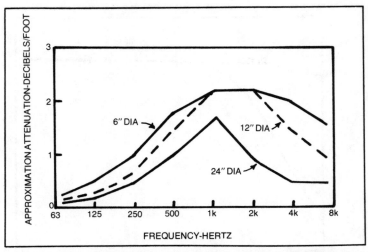

Fig. 12-7. Attenuation offered by spiral wound round ducts with perforated spiral wound steel liner. Dimensions shown are the free area inside the duct, no air flow. This is a plot of data selected from Table 19, Reference 92.

thickness of absorbent lining. A plenum located at the fan discharge can be an effective and economical way to decrease noise entering the duct system.

PACKAGED ATTENUATORS

Numerous proprietary packaged noise attenuators are available. Cross sections of several types are shown in Fig. 12-11 with their performance plotted below. For comparison, the attenuation of our old friend the lined duct is given in curve A. Some of the other attenuators have no line of sight through them, i.e., the

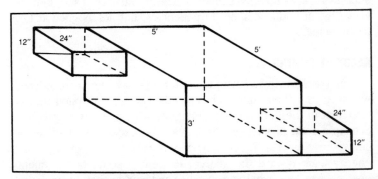

Fig. 12-8. Dimensions of a plenum which will yield about 21 dB attenuation throughout much of the audible spectrum.

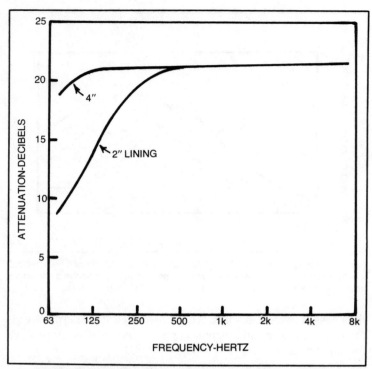

Fig. 12-9. Calculated attenuation characteristics of plenum of Fig. 12-8 lined with 2″ and 4″ glass fiber of 3 lb/cu ft density. Calculations follow ASHRAE procedures.

sound must undergo reflections from the absorbing material to traverse the unit and hence have somewhat greater attenuation. The absorbing material is usually protected by perforated metal sheets in these packaged silencers. The attenuation of such units is very high at midband speech frequencies, but not as good at low frequencies.

REACTIVE SILENCERS

Several passive, absorptive silencers have been considered which rely for their effectiveness on the changing of sound energy into heat in the interstices of fine glass fibers. Another effective principle used in silencers is that of the expansion chamber as shown in Fig. 12-12. This type performs by reflecting sound energy back toward the source, cancelling some of the sound energy. As there is both an entrance and exit discontinuity, sound is reflected from two points. Of course, the destructive interference (attenuation peaks) alternates with constructive interference (attenuation nulls) down through the frequency band, the attenuation peaks becoming lower as frequency is increased. These peaks are not harmonically related, therefore they would not offer high attenuation to a noise fundamental and all its harmonics, but rather attenuate slices of the spectrum. However, by tuning, the major peak can knock out the fundamental while most of the harmonics, of much lower amplitude, would receive some attenuation. By putting two reactive silencers of this type in series and tuning one to fill in the nulls of the other, continuous attenuation can be realized throughout a wide frequency range. No acoustical material is required in this type of silencer which operates like an automobile muffler.

RESONATOR SILENCERS

The resonator silencer, illustrated in Fig. 12-13, is a tuned stub which provides high attenuation at a single frequency or

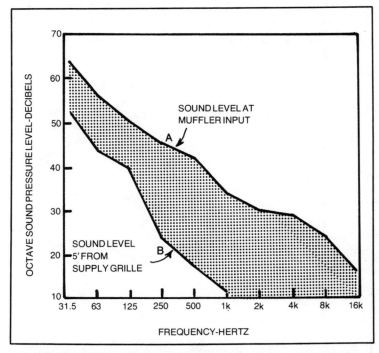

Fig. 12-10. Measured noise sound pressure levels (A) at plenum input and (B) in room 5 ft from supply grilles. This plenum is approximately the size of that of Fig. 12-8 but only half the height and it contains baffles.

narrow band of frequencies. Even a small unit of this type can produce 40 to 60 dB attenuation. This type of silencer offers little constriction to air flow which may be a problem with other types.

DUCT LOCATION

Why build a 60 dB wall between studio and control room, for instance, and then serve both rooms with the same supply and

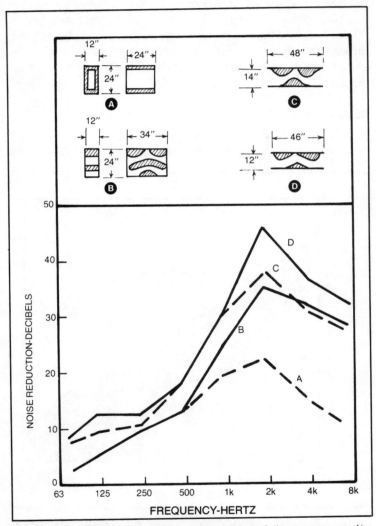

Fig. 12-11. Attenuation characteristics of three packaged silencers compared to that of the lined duct, curve A (adapted from Doelling, Reference 94).

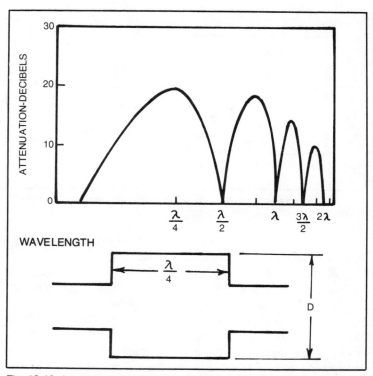

Fig. 12-12. Attenuation characteristics of reactive type of silencer, the expansion chamber. Sound is attenuated by virtue of the energy reflected back toward the source, cancelling some of the oncoming sound (adapted from Sanders, Reference 95).

exhaust grilles closely spaced as in Fig. 12-14? This is a tactical error which results in a short path speaking tube from one room to the other quite effectively nullifying the 60 dB wall. With an untrained air conditioning contractor doing the work without a supervisor sensitive to the acoustical problem, such errors can happen and do happen. To obtain as much as 60 dB attenuation in the duct system to match the construction of the wall requires application of many of the principles discussed above. Figure 12-15 suggests two approaches to the problem, to separate grilles as far as possible if they are fed by the same duct, or, better yet, to serve the two rooms with separate supply and exhaust ducts.

SOME PRACTICAL SUGGESTIONS

■ The most effective way of controlling air flow noise is to size the ducts to avoid high velocities. The economy of the smaller

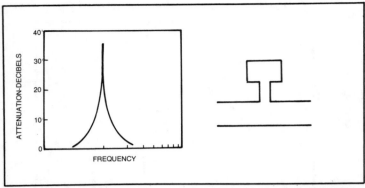

Fig. 12-13. Attenuation characteristics of the tuned stub silencer (adapted from Sanders, Reference 95).

ducts, however, may be more than enough to pay for silencers to bring the higher noise to tolerable levels.

■ Right angle bends, dampers, etc. create noise due to air turbulence. Locating such fittings 5 to 10 diameters upstream from the outlet allows the turbulence to smooth out.

■ Noise and turbulence inside a duct cause the duct walls to vibrate and radiate noise into surrounding areas. Rectangular ducts offend in this way more than round ones. Such noise increases with air velocity and duct size, but may be controlled with external treatment of thermal material.

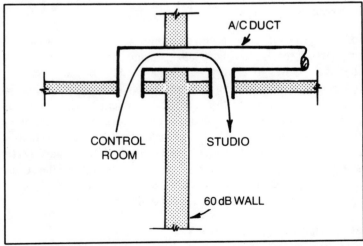

Fig. 12-14. An expensive, high transmission loss wall can be bypassed by sound traveling over a short duct path from grille to grille.

250

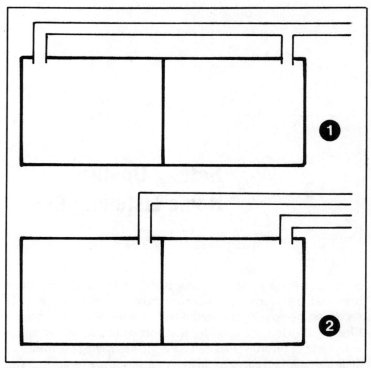

Fig. 12-15. Two possible solutions to the problem of Fig. 12-14, (1) separate grilles as far as possible, or (2) feed each room with a separate duct. Both approaches increase duct path length and thus duct attenuation.

■ Acoustical ceilings are not good sound barriers and in a sound sensitive area the space above a lay-in ceiling should not be used for high velocity terminal units.

■ The ear can detect sounds far below the prevailing NC contour noise (see Fig. 12-1). The goal should be to reduce noise in the studio to a level at which it cannot be heard on a playback of a tape recorder at normal level and without noise reduction.

■ Plenums are effective and straightforward devices adaptable to studio quieting programs and they offer attenuation throughout the audible spectrum. They are especially effective at the fan output.

■ Some of the noise energy is concentrated in the highs, some in the lows. There must be an overall balance in the application of silencers so that the resulting studio noise follows roughly the proper NC contour. Otherwise overdesign can result.

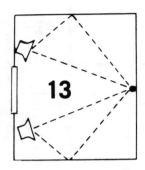

13 Setting Up the Home Listening Room

In the present context "listening room" is taken as the normal home hi-fi music center. The control room or monitoring room of a recording studio is a very special type of listening room treated in Chapter 16. Here we consider that portion of a home set up for enjoyment of recorded music. Those families having the means to dedicate a certain space exclusively for music are fortunate. The rest of us must "make do" with a multipurpose space, typically a living room serving also as a listening room.

Ideally, all members of the household will be of one mind in such a project. Realistically, there must be compromises between the technical and the esthetic. Just where the compromise point comes to rest (if ever) may turn out to be a measure of the relative force of personalities and strength of wills. If there is a genuine appreciation of good music shared by the persons involved, the war is over except for a few skirmishes along the way.

BINAURAL PERCEPTION

To a bystander a one-eyed person seems to function quite normally. Without a black patch, special eye glasses, or candystripe cane the blindness in one eye may not be noticed by others at all. The person having vision in only one eye certainly knows the difference. The ability to perceive depth and perspective and to estimate distance is seriously impaired.

The person who is deaf in one ear has a very similar problem. It is difficult to adjust to the inability to tell from which direction a sound comes. Background noises which with two good ears were taken in stride now become overpowering and harsh. Intelligibility of speech deteriorates and tiring concentration is constantly required just to get by. Also, stereo records and FM? Forget it. Everything is now in mono.

In Chapter 2 the marvelous marriage of form and function in the human ear was considered but, except for a brief discussion of binaural localization, the ear was considered as an isolated entity. But it really is not isolated. We have two ears and they function, along with the brain, as a binaural system. Just how signals from two ears are combined in the brain is not completely understood. Some think the two sets of signals are combined more or less equally in the brain. Another theory states that we actually listen with one ear at a time and that some sort of reciprocating suppression takes place. For instance, in listening with headphones which feed a signal to one ear differing from that to the other ear, alternating periods of dominance and suppression between the two ears is experienced. If congruous program material is fed to the two ears, true binaural hearing results as the two work harmoniously with the brain.

True binaural hearing gives us both localization of the sound source and a sense of acoustical perspective. It is possible to concentrate on a certain instrument in the orchestra, at the same time pushing all others into the background of consciousness. Or we may concentrate on the reproduced music to such an extent that we do not hear the front door bell ring.

THE ACOUSTICAL LINK

The acoustics of the space is a vital part of both the recording and reproducing process. In every acoustical event there is a sound source and some sort of receiving device with an acoustical link between the two. Disc or tape recordings have the imprint of the acoustics of the recording environment recorded on them. If the sound source is a symphony orchestra and the recording is made in the performing hall, the reverberation of the hall is very much a part of the orchestral sound. If the reverberation time of the hall is 2 seconds, a 2-second tail is evident on every impulsive sound and sudden cessation of music and it affects the fullness of all the music. In playing this record in our home listening area, what room characteristic will best complement this type of music?

Another recording may be of the rock and roll type. This music was probably recorded in a very dead studio by the multitrack system. Basic rhythm sections playing in this very dead studio and well separated acoustically were layed down on separate tracks. During subsequent sessions the vocals and other instruments were recorded on still other tracks. Finally, all were combined at appropriate levels in a mixdown, with a bit of "sweetening" added. The position in the stereo field of the sounds on each track is adjusted by turning a panpot knob. In the mixdown many effects, including artificial reverberation, were added. What listening room characteristics are best for playback of this recording?

If the taste of the hi-fi enthusiast is highly specialized, the listening room characteristics can be adjusted for relative optimum results for one type of music. If the taste is more universal, the acoustical treatment of the listening room may need adjustment for different types of music.

ELEMENTS OF SMALL ROOM ACOUSTICS

Small room acoustics are quite different from that of large halls or auditoriums because small room dimensions are comparable to the wavelength of the sound to be reproduced. The 20 Hz to 20 kHz frequency range covers sound wavelengths from 56.5 ft to 0.0565 ft (11/16 inch). Room resonances play a major role in determining the acoustics of a small room, especially below 300 Hz. We are thus forced to consider size and proportions after the fashion of Chapter 6.

The dynamic range of reproduced music in a listening room depends, at the loud extreme, on amplifier power, power handling ability of the loudspeakers, and the tolerance of family members and neighbors. The social limitation usually comes into play at a level far lower than the average electronic and transducer limitation. The soft end of the dynamic range scale is limited by noise, environmental or electronic. Household noise usually determines the lower limit. The usable range between these two extremes is far, far less than, say, the range of an orchestra in a concert hall, but we learn to be happy with what we have. Expanders, compressors, companders, etc., are capable of restoring some of the original range, but this is on a level of sophistication several notches above this discussion.

ROOM DIMENSIONS

Chapters 6 and 11 go into the problem of distributing the room modal frequencies. If our listening room is in the planning stage of a house yet to be built, room proportions and shape should receive the same attention as for a recording studio. Most of us look at the living room built a decade or two ago and say, "Well, it's this or nothing, what can be done?". The first thing to do is to find out how the room modes are distributed, following the steps indicated in Fig. 11-6. The situation will be close to ideal, just passable, or downright horrible. Rejoice if one of the first two, and be informed of potential problem frequencies if the latter.

Room size is all bound up with modal frequencies and their distribution. The smaller the room, the greater the spacing of room resonance frequencies. The old familiar BBC recommendation of a minimum volume of 1500 cu ft is as applicable to living rooms as to recording studios and control rooms.

ACOUSTICALLY COUPLED SPACES

Stairwells, entrance halls and doors opening into other rooms can affect the acoustics of the listening room. This effect is not necessarily all bad. If the coupled space is very live and the listening room dead, a double slope decay may result as energy from the live room feeds into the dead room, dominating the latter part of the decay. If the coupled spaces are treated more or less like the listening room, the effect may be helpful as sound is diffused during the rapid initial part of the decay.

REVERBERATION TIME

A good start in considering what reverberation time living rooms *should* have is what reverberation times they *do* have, on the average at least. Not many such studies have been made, but the results of two are shown in Fig. 13-1. The study of Jackson and Leventhall,[96] the upper graph, is the average of 50 British living rooms. The lower graph shows the results of a similar survey the British Broadcasting Corporation made in the living rooms of BBC engineers. Apparently the engineers' living rooms were better furnished, resulting in greater absorption and lower reverberation times. The BBC result at 125 Hz hints at the possibility of more frame dwellings among the engineers with resulting greater low frequency absorption as compared to apartments, many of which may be characterized by concrete and masonry, in the Jackson and Leventhall study. Insofar as the living rooms covered by these studies are similar to the specific living room being considered, we have a guide as to what reverberation time to expect without any further treatment.

The average reverberation times of Fig. 13-1 are really not far out of line for listening rooms, although quite low for listening to classical music. We must remember that reverberation time tells us nothing about diffusion of sound in the room or colorations that may exist. A range of reverberation time from 0.4 to 0.7 second is commonly accepted for home hi-fi listening rooms depending upon the type of program to be favored. In order for all frequency components of a musical signal to die away uniformly, the reverberation time should be reasonably uniform with frequency.

For organ music a long reverberation time is desirable, for the pizzicato plucked string number or staccato piano piece a shorter reverberation time may be desirable to avoid slurring of the fast flowing notes. In a practical sense, however, reverberation time must be adjusted to suit the individual taste, perhaps starting from some average point in the 0.4 to 0.7 second range and working on from this.

ADJUSTMENT OF REVERBERATION TIME

It is usually sufficient, for home listening room computations, to consider reverberation time at a single frequency. The frequency of 500 Hz is the one to pick because of its critical location in regard to human hearing and also concentration of signal energy. This simplifies things tremendously. First, the volume of the room should be calculated, then the total surface area of walls, floor, and ceiling. The Sabine equation is then recalled from Chapter 8:

$$RT60 = \frac{0.049\ V}{Sa} \qquad (13\text{-}1)$$

in which,

$RT60$ = reverberation time, seconds
V = volume of room, cu ft
S = surface area of room, sq ft
a = average absorption coefficient

Let us assume our living room is 14 x 25 x 8 ft in size. By cranking in a desired reverberation time, say, 0.5 second, a volume of 2800 cu ft and a total surface area of 1324 sq ft we get from Equation 13-1:

$$0.5 = \frac{(0.049)\ (2800)}{(1324)\ a}$$

$$a = 0.207$$

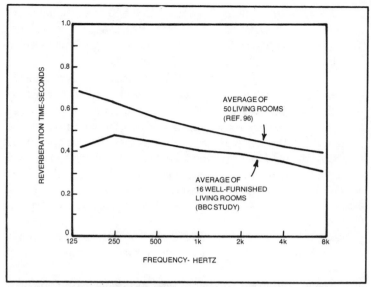

Fig. 13-1. Average reverberation time of 50 British living rooms as measured by Jackson and Leventhall and similar results they report from a study of the living rooms of 16 BBC engineers (after Jackson and Leventhall, Reference 96).

The total number of absorption units of the room is the average absorption coefficient of 0.207 multiplied by the total surface area of the room or (0.207) (1324) = 274 sabins. This 274 sabins establishes the absorption we *want*, but first we must estimate the absorption we *have*.

This takes a bit of guessing, but guessing (engineering estimates) is very much a part of acoustical treatment procedures. In the appendix is a list of the absorption coefficients of many materials for many frequencies, but to make this computation simple and self contained, Table 13-1 summarizes the 500 Hz coefficients for only the materials we need.

Let's figure the absorption units, piece by piece:

Wood floor:	(350 sq ft) (0.08) =	28.0 sabins
Gypsum board walls and ceiling:	(974 sq ft) (0.05) =	48.7
Carpet, 12 x 14 ft, heavy:	(168 sq ft) (0.30) =	50.4
Drapes, medium velour, draped to half area:	(84 sq ft) (0.49) =	41.2
		168.3 sabins

257

We have 105.7 sabins to go to reach the 274 sabins, but there are a davenport and two easy chairs, which appear to be very absorbent, as well as coffee table, end tables, and hard chairs which look like they may diffuse sound but not absorb much.

What does one do about a davenport? What area is exposed to sound? Mmmm, seven feet long, seat, back, and rear about six feet, that's 42 sq ft, and ends would add 12 sq ft, making about 54 sq ft total. This does not include the underside, but let's assume it is shielded from the sound. Here is the davenport area, how about the two easy chairs? Let's give them 20 sq ft each, bringing our total soft furniture area to 94 sq ft. What coefficient? With all that fabric and thick cotton padding, the absorption would certainly be better than heavy drapes folded to half area. Let's assume the absorption coefficient to be 0.7. This gives us $(94)(0.7) = 65.8$ more sabins, bringing our grand total to $168.3 + 65.8 = 234.1$ sabins. This is only 39.9 sabins short of what we need for a reverberation time of 0.5 second.

An interesting question along the way is what is the estimated reverberation time with 234.1 sabins? Cranking 234.1 sabins and volume into Equation 13-1 gives a reverberation time of 0.59 second. That is not too far from the 0.5 second goal. That 39.9 sabins might be reserved for doing some of the things we shall see need to be done in the next section.

PLACEMENT OF ABSORBING MATERIAL

Because the floor has the carpet and upholstered furniture, the vertical mode will be reasonably well damped, but the walls have only the gypsum board absorption, uniformly distributed. We have seen how patches of absorbing material not only are superefficient in absorption, but also diffusers of sound. We need a few patches to reduce the chance of N-S or E-W flutter echo. Here we run into a characteristic problem. The thought of hanging patches

Table 13-1. Selected Absorption Coefficients.

Material	Absorption Coefficient
Wood flooring	0.08
Gypsum board	0.05
Carpet, heavy	0.30
Drapes, medium velour draped to half area	0.49

of glass fiber on the walls of the living room is the kind of tinder that could start a conflagration. It will take some ingenuity to do this artistically. Drapery is a standby favorite. Bric-a-brac, cabinets, book shelves full of books, and large oil paintings provide that three dimensional requirement of effective diffusers. Artistic folding screens can have absorbent backs.

Having two unbroken reflective surfaces opposing each other, such as two walls, the possibility of repetitive echoes, called flutter echoes, exists. Flutter echoes are difficult to detect for close spacings, but with reflective walls spaced the length of larger rooms these echoes can be produced by snapping fingers or clapping hands. Even if not detectable by the above method, such echoes can give a roughness to sound in the room. It takes relatively little in the way of absorbing material or diffusing objects to discourage flutter echo formation.

ABSORPTION BY PEOPLE

Some of the 39.9 sabins of absorption will be supplied by people in the room enjoying the music. The Table in the appendix lists from 1.5 to 5.5 sabins absorption for each person at 500 Hz, depending on the softness of the seat they are sitting on. It becomes a bit complicated to balance the amount of davenport area a person shields by sitting on it against the person area added to the room, but it would seem reasonable to add 2 sabins per person while seated on our heavily upholstered seats. In Chapter 10 we saw that an informally dressed college student seated on a tablet arm chair absorbs 2.9 sabins at 500 Hz. This is consistent with the 2 sabin estimation and it allows 0.9 sabin for the davenport area covered.

ADJUSTABLE ACOUSTICS

One of the most acceptable absorbing devices for the living/listening room which has an adjustable capability is drapery material on a traverse rod. Figure 13-2 shows a drape 7 ft high and 12 ft wide. It has an area of 84 sq ft when closed and 28 sq ft when opened assuming 2 ft width on each side. With an assumed absorption coefficient of a = 0.49, this represents a change from 41 sabins to about 14 sabins, a 66% change. The percentage change is actually less than this because the coefficient for the deeply folded retracted drape would actually be somewhat higher than for the stretched out drape, but it gives us a rough estimation of the adjustments possible by this method. If the drape were retracted

Fig. 13-2. Drapes can provide needed absorption in the home listening room and provide an adjustable feature as well.

into a plywood receptacle, as in Fig. 13-3, the change would approach 100%.

Adjustable absorption can be accomplished in other ways such as panels supported by hooks and screw eyes, which can be introduced or removed from the room as desired. Hinged panels with one side reflective and the other side absorptive are effective, but most such methods are not well adapted to the usual home hi-fi room because of the lack of eye appeal.

The more meticulous audiophile may wish to incorporate some form of adjustable absorption to optimize conditions for the type of music to be listened to or to compensate for the number of persons in the room.

THE NIPPON GAKKI EXPERIMENTS

Considerable energy has been expended in trying to unravel some of the acoustical mysteries of specialized rooms such as studios and music halls. The home music room is, by its very nature, such an indefinite and highly variable space as to discourage research on it. A recent investigation[97] at the Nippon Gakki Co., Ltd. of Japan was actually directed toward the design of a listening room for evaluating audio products by users, dealers, and audio professionals. The results, however, can be applied to home listening rooms.

Very briefly, the test rooms were equipped with parquet wood floors and absorptive ceilings and with drapes and absorbing material distributed to suit the needs of each test. The reverberation times of the test rooms were in the neighborhood of 0.3 second. Four audio professionals, listening to different types of vocal and instrumental music, made judgements on (1) localization, (2) coloration, (3) perspective, (4) loudness, and (5) broadening of sound image.

At the risk of failing to do justice to a very complex and interesting paper, some of the results are summarized. For critical monitoring and evaluation of audio products it is best to absorb early reflections from walls, but for fully enjoying the music, early reflections were found to be important. Better localization of the sound image was found with an absorptive wall behind the

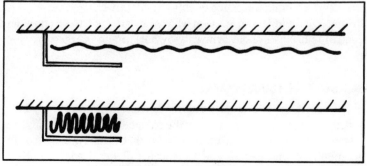

Fig. 13-3. The range of adjustment of drape adjustment can be increased by retracting the drapes into a covered receptacle.

loudspeakers. Broadening of the sound image results with reflective side walls. Colorations (change in tonal quality) were observed in all test arrangements.

Highly detailed and sophisticated measuring techniques yielded some interesting objective confirmation of the subjective judgements. With reflective side walls it was found that the left wall reflection was 8 dB down from the direct and delayed 4 milliseconds. For the right wall the corresponding figures are 10 dB down and 8.4 milliseconds delay. With absorptive side walls, wall reflections were down another 5 to 7 dB. Utilizing cross correlation techniques which measure the similarity of direct and reflected components they reported a coefficient of 0.28 for the reflective walls and 0.44 for the absorbing walls. This explains the superiority of the "tight" condition with absorptive walls in evaluating audio products. It seems that even though reflections from highly reflective walls bear less resemblance to the direct sound, their contribution to enjoyment of music is positive and significant.

As noted above, colorations, or alterations of tonal quality, were noted in all test arrangements. Certainly with both side walls reflective, as it was in some tests, a full-blown lateral room resonance would exist. At the fundamental frequency of about 30 Hz corresponding to the width of the room, there would be a sound pressure null midway between the walls and the same would be true for 90 Hz, 150 Hz, etc. For the even multiples (60, 120, 180 Hz, etc.) there would be a pressure peak at the same point.

Such effects account for the colorations, but are colorations all bad? We should not consider them a fault because they are so much a part of our everyday life. We like to listen to music in rooms, and all rooms have different response at different positions in the room. No hope for eliminating such effects completely can be offered, but the good listening room is one in which they are controlled and such control is brought about primarily by adjustment of room proportions.

BALANCING OF LOUDSPEAKERS

Even if the electronics and loudspeakers are perfectly balanced, room acoustics may upset the balance. For example, if such a balanced system is placed in a room that is unsymmetrical in geometry or one having one side wall reflective and the other absorptive, the loudspeaker balanced as judged at a given location may be affected. Cook has suggested a novel method of evaluating such acoustical effects on left-right stereo perception.[98] He has reached back into early aviation history and resurrected the old N-A aircraft navigation system. In this system one radio transmitter near the landing field transmits the letter A (dot-dash) while a second nearby transmitter synchronized with the first, transmits the letter N (dash-dot). When the airplane is perfectly on course the Ns and As merge to give a constant tone. If the plane is a trifle off course on one side the N would begin to be heard; if far off course the N strongly dominates. If the airplane drifts to the other side of the intended course, the A dominates.

By rigging a device that interrupts ⅓ octave random noise bands so that an A comes out at one output and N out the other, the N signal is fed to the left channel and the A to the right channel. At a given point in the listening area of the room it can readily be determined whether the A or N loudspeaker dominates and the degree can be estimated. Perceiving a steady ⅓ octave noise means that electronics, loudspeakers, room acoustics, *and ears* are in perfect balance for that spot *and for that frequency*. The ⅓ octave noise signal is about as narrow as can be used and still retain the important statistical effect. Such bands of noise avoid the disconcerting wild fluctuations of a pure tone. Cook reports that the range 200 to 3000 Hz yields the more useful localization data. Components below 200 Hz contribute misleading information resulting from reflections.

TESTING THE ROOM

Few audio buffs are equipped with sophisticated electronic measuring equipment to separate out different reflected components and study the intensity, delay, and spectrum of each. Few will be even able to measure reverberation time, but this can be estimated by calculations. However, those who have a sound level meter (even the cheapies will do a fair job) or can borrow one can measure the acoustic response of the room. There are numerous phonograph discs on the market[99] having ⅓ octave bands of pink noise. By reproducing these and measuring the band level at a certain point in the listening area, the overall response can be measured which includes the response of the pickup, amplifiers, loudspeaker, and the acoustics of the space. It is then possible to do a modest amount of equalization to mold the response to the desired shape. It must be remembered that it is not possible to equalize out basic acoustical flaws of the room. If boominess is encountered, reducing response at low frequencies may be helpful in controlling it.

SUMMARY

It is futile to idealize the home listening environment in this discussion because most home listening rooms are far from ideal as they serve multiple purposes. Not all of the suggestions below can be followed in many cases, but they can serve as guides to music room improvement.

Reverberation. Avoid excessively long or very short reverberation time and try to achieve something in the range of 0.4 to 0.7 second and adjust to suit taste in music. If the room is of frame construction do not worry about bass rise of reverberation time. If masonry, some low frequency absorption may be required to avoid a "boomy" effect (Chapter 10).

Room Resonances. In new construction adjust room proportions to distribute the resonance frequencies (Chapter 6) properly. For existing spaces learn to live with room resonances, but with an understanding of the causes of coloration and, possibly, identification of the offending frequencies.

Sound Diffusion. Room resonances are the cause of gross deviations from a diffuse state. Control flutter echoes between opposing, parallel, reflective walls by placement of absorbers or diffusing objects or by inclined reflective planes.

Noise. Isolating the music room from rooms that are noise sources and from bedrooms and other quiet areas is an ideal, at least.

Materials. The treatment of room surfaces may draw from the following:

■ Rugs, drapes, carpets which absorb primarily at high frequencies.

■ Acoustical tiles and other proprietary materials usually having high midband absorption. Low frequency absorption can often be enhanced by mounting so that there is an air space behind the material.

■ Plywood paneling (not the ersatz thin paneling cemented to an existing wall) is attractive and offers modest low frequency absorption with air space behind. Paneling may be nicely balanced with drapes.

■ Frame construction usually gives sufficient low frequency absorption. Masonry construction usually gives insignificant absorption, lows or highs.

■ Concrete block walls are highly variable in their absorption. If light and porous, their absorption may be quite high if unpainted or painted with a non-bridging paint. A paint which seals the interstices reduces sound absorption to a very low value.

■ Wood absorbs primarily at high frequencies and unpainted wood is a better absorber than painted. This applies to surface absorption. As a floor diaphragm, good low frequency absorption results.

■ Sound absorption in the air is negligible for home sized rooms.

Snow[100] made a good point in comparing film production to listening to music. In a film the story is the important thing and technique will never compensate for a mediocre story. In the home the music is the important thing and made to be enjoyed. Perhaps you will be able to utilize some of the factors discussed in this chapter to improve your enjoyment, but never let them dominate to the detriment of the enjoyment of the music. Two other helpful papers on the subject of this chapter are listed in the references.[101,102]

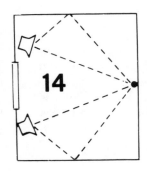

14 Building a Studio

What is a *good* recording studio? There is only one ultimate criterion—acceptability of product. In a commercial sense a successful recording studio is one fully booked and making money. Music recorded in a studio is pressed on discs or recorded on tape and sold to the public. If the public likes the music, the studio passes the supreme test. There are many factors influencing the sale of the recording beside studio quality, such as type of music and popularity of the musicians, but studio quality is vital, at least for success on a substantial, long range basis.

There are hundreds of small recording studios operated usually by not-for-profit organizations which turn out a prodigious quantity of material for education, promotional, and religious purposes. With limited budgets and limited technical resources the operators of such recording studios are often caught between a desire for top quality and the lack of means and, often, the know-how to achieve it. To this group could also be added the musician or advanced audiophile wanting recording facilities for personal development or to explore future commercial possibilities. This chapter is aimed primarily to those in this needy group although the principles are more widely applicable.

Public taste must be pleased for any studio to be a success. Producing a successful product, however, involves many individuals along the way whose decisions may make or break a studio. These decisions may be influenced by both subjective and technical factors. The appearance of a studio, convenience, and comfort may outweigh acoustical quality, sometimes because the more tangible esthetic qualities are better understood than the intangible acoustical qualities. This chapter has little to say on the artistic, architectural, and other such aspects of a studio, but their importance cannot be denied. They just require a different brand of specialist.

LEARNING TO PLAY A STUDIO

Sound picked up by a microphone in a studio consists of both direct and indirect sound. The direct sound is the same as that which would exist in the great outdoors or in an anechoic chamber. The indirect sound, which immediately follows the direct, is that sound which results from all the various non-free field effects characteristic of an enclosed space. This latter is unique to a particular room and may be called the studio response. Another way to look at it is that everything that is not direct sound is indirect sound which can be lumped together as reverberation.

Before we dissect indirect sound, let us look at it in its all inclusive reverberatory form in a studio, or any other room for that matter. Figure 14-1 shows how sound level varies with distance from a source S which could be the mouth of one talking, a musical instrument, or a loudspeaker. A sound pressure level of 80 dB is measured 1 ft from the source. If all surfaces of the room were 100% reflective, we would have a reverberation chamber to end all reverberation chambers and the sound pressure level would be 80 dB everywhere in the room because no sound energy is being absorbed. There is essentially no direct sound, it is all indirect. Graph B represents the fall-off in sound pressure level as one moves away from the sound source with all surfaces 100% absorptive. In this case all the sound is direct, there is no indirect. The best anechoic rooms approach this condition, but never completely. It is the true free field illustrated in Figs. 4-1 and 4-2 and for this condition the sound pressure level decreases 6 dB for each doubling of the distance.

Between the indirect "all reverberation" case of graph A of Fig. 14-1 and the direct "no reverberation" case of graph B lies a multitude of other possible "some reverberation" cases, depending on room treatment. In the area between these two extremes lies the real world of studios where we live and move and have our being. The room represented by graph C is much more "dead" than that of graph D. In practical studios the direct sound is observable out a short distance from the source but after that the indirect sound dominates. A sudden sound picked up by a microphone in a studio would, for the first few milliseconds, be dominated by the direct component after which the indirect sound arrives at the microphone as a torrent of reflections from room surfaces. These are spread out in time because of the different path length traveled.

A second component of indirect sound results from room resonances, which in turn are the result of reflected sound. The direct sound flowing out from the source excites these resonances, bringing into play all the effects listed in Chapter 6. When the source excitation ceases, each mode dies away at its own rate. Sounds of very short duration may not last long enough to fully excite room resonances.

The third component of indirect sound is involved with the materials of construction - doors, windows, walls, floors. These, too, are set into vibration by sound from the source and they, too, decay at their own particular rate when excitation is removed. If Helmholtz resonators are involved in room treatment, we recall that sound not absorbed is reradiated.

The sound of the studio, embracing these three components of indirect sound plus the direct sound, has its counterpart in musical instruments. In fact, it is helpful to consider our studio as an instrument which the knowledgeable musician, technician, or engineer can play. It has its own characteristic sound and skill is required to extract from it its full potential.

Reverberation is the composite, average effect of all three types of indirect sound. Measuring reverberation time does not get at the individual components of which reverberation is constituted. Herein lies the weakness of reverberation time as an indicator of studio acoustical quality. The important action of one or more of the indirect components may be obscured by the averaging process. This is why it is said that reverberation time is *an* important indicator of studio acoustical conditions, but not the only one.

STUDIO DESIGN

In a general book of this type, space is too limited to go into anything but basic principles. Fortunately, there is a rich literature on the subject, much of it written in easy-to-understand language. In designing a studio, attention should be given to room volume, room proportions, reverberation time, diffusion, and isolation from interfering noise.

STUDIO VOLUME

A small room almost guarantees sound colorations resulting from excessive spacing of room resonance frequencies. This can be

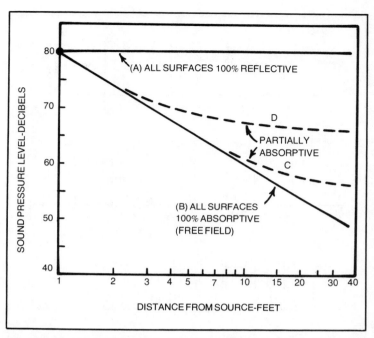

Fig. 14-1. Variation of sound pressure level in an enclosed space as distance from the source of sound is increased depends on the absorbency of the space.

illustrated by picking one of the favorable room ratios suggested by Sepmeyer,[44] 1.00 : 1.28 : 1.54, and applying it to a small, a medium, and a large studio and see what happens. Table 14-1 shows the selected dimensions, based on ceiling heights of 8, 12, and 16 ft resulting in room volumes of 1000, 3400, and 8000 cu ft. Axial mode frequencies were then calculated after the manner of Fig. 11-6 and plotted in Fig. 14-2, all to the same frequency scale. As previously noted, the room proportions selected do not yield perfect distribution of modal frequencies, but this is of no consequence in our investigation of the effects of room volume. A visual

Table 14-1. Studio Dimensions.

	Ratio	Small Studio	Medium Studio	Large Studio
Height	1.00	8.00 ft	12.00 ft	16.00 ft
Width	1.28	10.24 ft	15.36 ft	20.48 ft
Length	1.54	12.32 ft	18.48 ft	24.64 ft
Volume		1000 cu ft	3400 cu ft	8000 cu ft

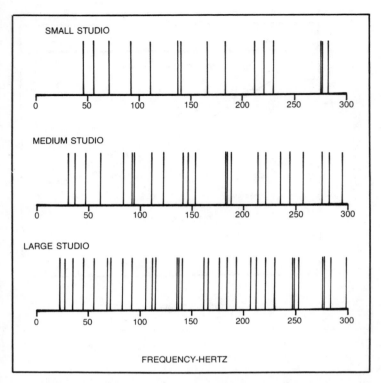

Fig. 14-2. Comparison of the axial mode resonances of a small (1000 cu ft), a medium (3400 cu ft), and a large (8000 cu ft) studio, all having 1.00:1.28: 1.54 proportions.

inspection of Fig. 14-2 shows the increase in the number of axial modes as volume is increased which, of course, results in closer spacing. In Table 14-2 the number of axial modes below 300 Hz is shown to vary from 18 for the small studio to 33 for the large. The low frequency response of the large studio is shown to be far superior to that of the two smaller studios, 22.9 as compared to 45.9 Hz. This is an especially important factor in the recording of music.

We must remember that modes other than axial are present. The major diagonal dimension of a room better represents the lowest frequency supported by room resonances because of the oblique modes. Thus the frequency corresponding to the room diagonal listed in Table 14-2 is a better measure of the low frequency capability of a room than the lowest axial frequency. This approach gives the lowest (gravest) frequency for the large room as 15.8 Hz compared to 22.9 Hz for the lowest axial mode.

The average spacing of modes, based on the frequency range from the lowest axial mode to 300 Hz, is also listed in Table 14-2. The average spacing varies from 8.4 Hz for the large studio to 14.1 Hz for the small.

The reverberation times listed in Table 14-2 are assumed, nominal values judged fitting for the respective studio sizes. On the basis of these reverberation times, the mode bandwidth is estimated from the expression 2.2/RT60. Mode bandwidth varies from 3 Hz for the large studio to 7 Hz for the small studio. The advantage of closer spacing of axial modes in the large studio tends to be offset by its narrower mode bandwidth. So, we see conflicting factors at work as we realize the advantage of the mode skirts overlapping each other. In general, however, the greater number of axial modes for the large studio, coupled with the extension of room response in the low frequencies, results in its superiority over the small studio.

The examples of the three hypothetical studios considered above further emphasize the analogy of a musical instrument and a studio. We can imagine the studio as a stringed instrument, one string for each modal frequency. These strings respond sympathetically to sound in the room. If there are enough strings tuned to closely spaced frequencies and each string responds to a wide enough band of frequencies to bridge the gaps between strings, the studio-instrument responds uniformly to all frequency components of the sound in the studio. If the lines of Fig. 14-2 are imagined to be strings, it is evident that there will be dips in response between widely spaced frequencies. The large studio, with many strings,

Table 14-2. Studio Resonances in Hertz.

	Small Studio	Medium Studio	Large Studio
Number of axial modes below 300 Hz	18	26	33
Lowest axial mode	45.9	30.6	22.9
Average mode spacing	14.1	10.4	8.4
Frequency corresp to room diagonal,	31.6	21.0	15.8
Assumed reverb, time of studio, second	0.3	0.5	0.7
Mode bandwidth (2.2/RT60)	7.3	4.4	3.1

yields the smoother response. Conclusion: A studio having a very small volume has fundamental response problems in regard to room resonances; greater studio volume yields smoother response. The recommendation based on BBC experience still holds true, that coloration problems encountered in studios having volumes less than 1500 cu ft are such as to make such small rooms impractical. The axial modes considered in the above discussion for reasons of simplicity are not the only modes, but they are the dominant ones.

ROOM PROPORTIONS

If there are fewer axial modes than one would desire in the room under consideration, sound quality is best served by distributing them as uniformly as possibly. The cubical room distributes modal frequencies in the worst possible way: piling up all three fundamentals, and each trio of harmonics with maximum gap between. Having any two dimensions in multiple relationship shares in this type of problem. For example, a height of 8 ft and a width of 16 ft means that the second harmonic of 16 ft coincides with the fundamental of 8 ft. This emphasizes the desirability of proportioning the room for best distribution of axial modes.

The perfect room proportions have yet to be found. It is easy to place undue emphasis on a mechanical factor such as this. The reader is urged to be informed on this subject of room resonances and to be aware of certain consequences, but let us be realistic about it - all of the recording that has ever taken place has been done in spaces less than perfect. In our homes and offices conversations are constantly taking place with serious voice colorations and we listen to and enjoy recorded music in acoustically abominable spaces. The point is that in striving to upgrade sound quality at every stage of the process, reducing sound colorations by attenuation to room modes is just good sense.

REVERBERATION TIME

If reverberation time is too long, speech syllables and music phrases are slurred and a definite deterioration of speech intelligibility and music enjoyment results. If rooms are too dead, i.e., reverberation time too short, speech and music lose character and suffer in quality, music suffering the greater amount. These effects are not so definite and precise as to encourage thinking that there is a specific optimum reverberation time because many other factors are involved. Is it a male or female voice, slow or fast talker,

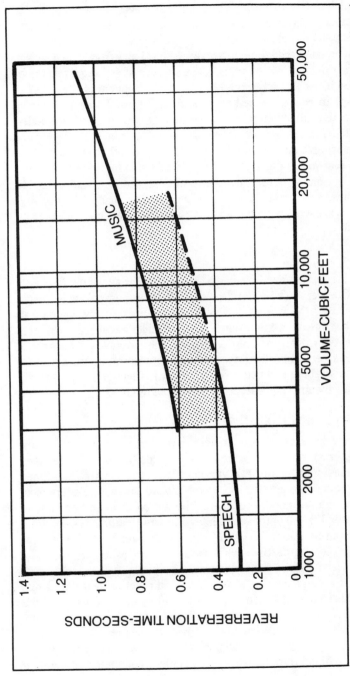

Fig. 14-3. Suggested reverberation times for recording studios. The shaded area is a compromise region for studios in which both music and speech are to be recorded.

273

English or German (they differ in the average number of syllables per minute), a standup comic or a string ensemble, vocal or instrumental, hard rock or a waltz? In spite of so many variables, readers need guidance and there is a body of experience from which we can extract helpful information. In the process of writing a book published in 1973, I attempted to distill the "wisdom of the ages" in regard to recommended reverberation times. I found scores of subjective judgements, not too consistent, which have been boiled down to Fig. 14-3, taken from that book.[103] Active in designing and measuring studios during the intervening period, I find no persuasive reason for altering it. It should be taken for what it is, not a true optimum, but following it will result in reasonable, usable conditions for many types of recording. The shaded area of Fig. 14-3 represents a compromise in rooms used for both speech and music.

NOISE

It is emphasized again that noise is in the ear of the "behearer". One person's beautiful music is another person's noise, especially at 2 A.M. It is a two way street and, fortunately, a good wall which protects a studio area from exterior noise also protects neighbors from what goes on inside. The psychological aspect of noise is very important; acceptable if considered a part of a situation, disturbing if considered extraneous. Chapter 12 has already treated the special case of air conditioning noise.

STUDIO DESIGN LITERATURE

We have considered reverberation and how to compute it (Chapter 8), the reality of room resonances (Chapter 6), the need for diffusion (Chapter 11), various types of dissipative and tuned absorbers (Chapter 10), and, as mentioned above, one of the most serious studio noise producers, the air conditioning equipment (Chapter 12). All of these are integral parts of studio design. What more can be done? Well, the would-be designer should at least sample the very pertinent published books and papers to see how others have solved their problems.

Bruce's paper[104] is a whimsical and practical account of one who has tread the studio design path. Dilley's article[105] tells of the transformation of a frozen food locker into a successful studio. Penner's paper[106] is unusual in that it recounts the research, thought processes, and decisions along the way to make a studio out of an Illinois farmhouse, and it includes a complete bibliography

Fig. 14-4. View of 2500 cu ft voice studio looking into the control room (courtesy of World Vision, International).

of 134 entries covering general books on acoustics, reverberation, measurements, and studio design and construction. The Storyk and Wolsch article [107] traces the problems the designers encountered in three different studios. The six part series by Rettinger[108]

Fig. 14-5. Rear view of the 2500 cu ft voice studio of Fig. 14-4. Wall modules containing 4 inches of dense glass fiber contribute to diffusion of sound in the room (courtesy of World Vision, International).

Fig. 14-6. A 3400 cu ft studio used for both recording voice and editing tapes. Decorator type of fabric makes visual feature out of the absorber/diffuser wall modules (courtesy of Mission Communications Incorporated).

gives a qualified acoustical consultant's approach to numerous studio problems and is highly recommended. The author's book[109] gives fairly complete designs for a dozen budget studios.

SOME STUDIO FEATURES

A glance into other people's studios stimulates ideas such as "things I want to do" or "things I definitely don't like". Figures 14-4 and 14-5 show the treatment of a budget 2500 cu ft studio. Built on the second floor of a concrete building with an extensive printing operation below, certain minimum precautions were advisable. The studio floor is ¾" plywood on 2 x 2 inch stringers resting on ½" soft fiberboard. Attenuation of noise through the double ⅝" drywall ceiling is augmented by a one inch layer of dry sand, a cheap way to get amorphous mass. The wall modules, containing a 4" thickness of Owens-Corning Type 723 Fiberglas (3 lb/cu ft density), help to diffuse the sound.

The studio of Fig. 14-6, with a volume of 3400 cu ft, has a couple of interesting features. The wall modules, Fig. 14-5, feature carefully stained and varnished frames and a neat grillecloth. Those of Fig. 14-7 are of two kinds, one sporting a very attractive fabric design, the other more subdued. The studio of Fig. 14-6 has a rather high ceiling, hence a virtual, visual ceiling was established

276

at an 8 ft height. This consists of four 5 x 7 ft suspended frames as shown in Fig. 14-8 which hold fluorescent lighting fixtures and patches of glass fiber. The plastic louvers are acoustically transparent.

Fig. 14-7. Close view of wall modules in studio of Fig. 14-6 (courtesy of Mission Communications Incorporated).

Fig. 14-8. The high structural ceiling of the studio of Fig. 14-6 allows the use of four 5 x 7 ft suspended frames to bring the visual ceiling down to 8 ft and to support illumination fixtures and absorbing material. The plastic louvers are acoustically transparent (courtesy of Mission Communications Incorporated).

The voice studio of Fig. 14-9, having a volume of 1600 cu ft, employs wall absorbing panels manufactured by the L.E. Carpenter Co. of Wharton, New Jersey. These panels feature a perforated vinyl wrapping and a ⅜" rigid composition board backing.

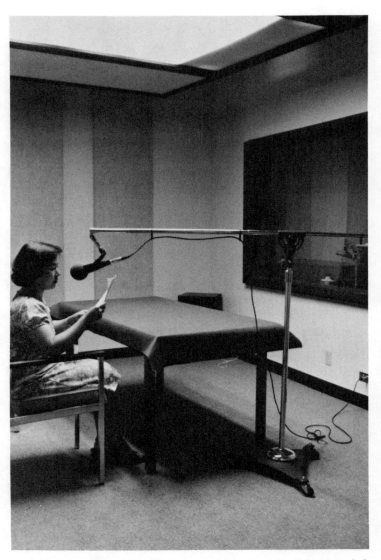

Fig. 14-9. Voice studio having a volume of 1600 cu ft. A 7 x 10 ft suspended ceiling frame hides 13 Helmholtz resonators for low frequency absorption. Wall modules are proprietary units covered with perforated vinyl (courtesy of Far East Broadcasting Company).

The concrete floor rests on soft fiberboard with distributed cork chips under it. The low frequency deficiencies of carpet and wall panels require some Helmholtz correction and thirteen 20 x 40 x 8 inch boxes are mounted in the suspended ceiling frame out of sight.

Fig. 14-10. Music studio of 3700 cu ft volume employing two 9 x 11 ft suspended ceiling frames which hold a total of 28 Helmholtz resonators (courtesy of Far East Broadcasting Company).

Figure 14-10 is a 3700 cu ft music studio which is also used for voice work. Low frequency compensation is accomplished by the same Helmholtz boxes mentioned above, fourteen of them in each of two suspended ceiling frames.

ELEMENTS COMMON TO ALL STUDIOS

Chapter 2 of Reference 109 treats sound lock treatment, doors and their sealing, wall constructions, floor/ceiling constructions, wiring precautions, illuminating fixtures, observation windows, and other such things which are common to all studios and which can give serious problems if not handled properly.

Multitrack Recording — A Special Case

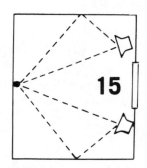

In the very early days of recording, artists crowded around a horn leading to a diaphragm-driven stylus cutting a groove on a wax cylinder. In early radio dramatics, actors, actresses, and sound effects persons moved in toward the microphone or faded back according to the dictates of the script. Greater freedom came as several microphones, each under separate control in the booth, were used. All this, of course, was monophonic.

Monophonic is still with us in commercial form, principally in AM radio, but the advent of stereophonic techniques has added a new dimension in realism and enjoyment of recordings, films, FM radio, etc. Stereo requires, basically, a dual pickup. This may be two separated microphones or two elements with special directional characteristics and electrical networks mounted close together. In broadcasting or recording a symphony orchestra, for example, it was soon found that some of the weaker instruments required their own microphones to compete with louder instruments. The signal from these was proportioned between the left and right channels to place them properly in the stereo field. Here, again, we see a trend from two microphones to many.

Popular music has always been with us, but its form changes with time. Recording techniques came to a technological maturity just in time to be clasped to the breast of new wave musicians and musical directors. Whether the Beatles were truly the vanguard of this new musical development will be left to historians, but their style spread like wildfire throughout the western world. "Good" sound quality in the traditional sense was not as much sought after as was a distinctive sound. Novelty effects such as phasing and flanging sold records by the millions. A new era of studio recording, variously called *multichannel, multitrack*, or *separation* recording, burst upon the scene. It was beautifully adapted to the production of special effects and the novel, distinctive sound, and it flourished.

FLEXIBILITY

The key word is "flexibility". Multitrack provides the means for recording one instrument or soloist of a group at a time, if desired, as well as the introduction of special effects along the way. A production can be built up piece by piece and assembled later in the mixdown. In Fig. 15-1A the signals from several microphones are combined in a summing network and fed to a single track recorder. A variation is to use a two-track recorder, distributing signals from each microphone, partially or wholly, between the two tracks for stereophonic recording and reproduction. In contrast, the signal of each microphone of Fig. 15-1B is recorded on a single track of a multitrack recorder. Many variations of this arrangement are possible. For example, a half dozen microphones on the drums could be premixed and recorded on a single track, but mixdown flexibility would be sacrificed in the process.

Once all the component parts of a musical production are recorded synchronously on separate tracks, they can then be mixed down to mono, stereo, or quadraphonic form for release. Much attention can be lavished on each detail in the mixdown, a stage which becomes a very important part in the production chain of events.

ADVANTAGES OF MULTITRACK

As mentioned above, flexibility is the outstanding overall advantage of multitrack techniques, but to understand the true breadth of the word some supporting detail is offered. Multitrack recording makes possible the conquering of space and time. Let us say that the drums and electric piano are recorded on separate tracks on Monday. On Tuesday the guitar player is available so a third track is recorded as he listens to a temporary mix of the first two on headphones. The tape may be shipped across the country to another studio to pick up a big name female vocalist between engagements. In this way a musical production can be built up a

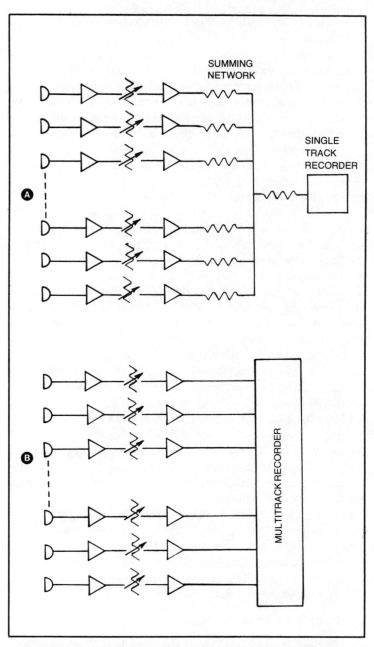

Fig. 15-1. The output from several microphones can be combined by a summing network and recorded on a single track tape (A). In multitrack recording (B) the output from each microphone is recorded on a separate track of the tape.

piece at a time. Perhaps this is not the best way, but it is a possibility.

Another big advantage of separation recording is the element of almost complete control over relative levels of each instrument and artist. Each track can be given just the equalization desired, often after considerable experimentation. Special effects can be injected at the mixdown stage. Reverberation can be added from reverberation room, plate, or spring in any desired amounts.

If perfection is the goal, it is an expensive commodity when artists, musicians, and technical crews are standing by for a retake of one performer or to argue the next step. The mixdown is a calmer session than recording a group on a premix basis.

There is also a noise advantage in multitrack. In a "mix and record as the group plays" type of session, levels of the various instruments are adjusted and the result, as recorded, is frozen as the final mix with no recourse. Some potentiometers are set high and some low depending on the requirements of the source and the signal to noise ratio of each channel is thereby fixed. In separation type recording the standard practice is to record all channels at maximum level which guarantees the best signal to noise ratio on the original tape. In the multichannel mixdown these levels are adjusted downward to achieve the desired balance, but there still remains a significant noise advantage over the premixed case.

In addition to recording each track at maximum undistorted level, the bandwidth of some channels can be reduced without affecting the sound of the instruments. For example, violin sound has practically no energy below 100 Hz. By chopping the low frequency end of that channel at 100 Hz, the noise is reduced with no noticeable degradation of quality. The sound of some instruments is essentially unaltered by cutting some low frequency energy, some by cutting highs.

A pervasive argument for multichannel recording is that it is what the client wants, expects, and gets. Customer demand, in the final analysis, may be the greatest factor influencing the drift toward multichannel. Of course, the customer demands it because of the inherent flexibility, potential savings, and other virtues, so we are back where we started.

DISADVANTAGES OF MULTITRACK

In spite of the advantage of signal to noise ratio mentioned previously, multitrack carries a disadvantage of noise buildup as the number of tracks combined is increased. When two tracks having equal noise levels are mixed together, the noise on the combined track is 3 dB higher than either original. If 32 tracks are engaged in a mixdown, the combined noise is 15 dB higher than a single track. Table 15-1 lists noise buildup for commonly used track configurations. It is simply a matter of adding noise powers. Mixing eight tracks of equal noise powers means that the total noise is $10 \log 8 = 9.03$ dB higher than the noise of one track. If the noise of one track is -80 dB referred to the reference level, the noise of 16 tracks would be -68 dB.

Dynamic range of a system is defined as the total usable range of audio level between the noise at the lower extreme and the full undistorted level at the upper extreme. The more tracks on a given width of tape, the narrower each track and the lower the reproduced signal level. Increased noise and decreased reproduced level spell narrower dynamic range.

The closer the spacing of tracks on a tape the greater the cross talk between adjacent tracks. Recording circumstances determine the magnitude of the resulting problem. For instance, if the two adjacent tracks are of two musical instruments recorded simultaneously in the studio while playing the same number, the congruity may make the crosstalk acceptable. The degree of separation realized between microphones in the studio affects the judgement on the seriousness of tape cross talk. If the material on adjacent tracks is unrelated (usually not the case in music recording) the cross talk will, of course, be much more noticeable.

Artistic responsibility may become diffused in multitrack recording unless the musical director is intimately involved in both recording and mixdown. The very nature of the mixdown technique dictates seemingly endless hours of detailed comparison of tracks, recording pickups, and overdubs which are the basic creative steps in a production. Often this meticulous duty falls on the recording

Table 15-1. Multitrack Noise Buildup.

Number of Tracks	Noise Buildup Above Noise of One Track, Decibels
2	3.01
4	6.02
8	9.03
16	12.04
24	13.80
32	15.05

engineer with only an occasional check by the music director. In contrast, the old style of premix recording session ends with an essentially completed product with the music director in full charge all the way.

Some separation recording sessions tend to separate musicians to the extent of losing some spontaneous interaction. Musicians respond to each other and this desirable effect may or may not be maintained in the face of cueing by foldback headphones and as the musicians are physically isolated by baffles, screens, and isolation booths.

While we are considering the negative aspects of multitrack recording, the degradation of quality as the tape is run and rerun scores of times must be added to the rest. In what other endeavor does the original recording receive such treatment? With 2 inch magnetic tape, contact with the head becomes a problem as the tape is passed through the machine, especially the outer tracks. The wise recording engineer reserves for outside tracks those sounds least affected by loss in high frequency response.

ACHIEVING TRACK SEPARATION

Achieving 15 to 20 dB intertrack separation requires intelligent effort and attention to detail. Without such separation the freedom of establishing relative dominance in the mixdown is sacrificed. The following methods are employed to yield the required separation:

—Adjustment of acoustics of studio
—Spacing the artists
—Using microphone placement and directivity
—Use of physical barriers
—Use of gating techniques
—Use of contact transducers

STUDIO ACOUSTICS

Heretofore in considering studio acoustics the criterion has been quality (naturalness and freedom from colorations) of the recorded signal. For multitrack recording emphasis is shifted to track separation and the very word "quality" tends to be replaced by "distinctive sound" in the process. Reflective surfaces in the studio are not ruled out, but they are generally localized for specific instruments while the general studio acoustics are made quite dead and absorptive. The number of musicians to be accomodated is limited, among other things, by the size of the studio. If walls are highly absorptive, musicians can be placed closer to them and more artists can be accomodated in a given space. Reverberation time rather loses its meaning in a studio specializing in separation recording, but, if measured, it would tend to be quite short.

DISTANCE BETWEEN ARTISTS

With an absorptive studio, increasing distance between the various instruments is a step toward track separation. Sound level falls off at a rate of 6 dB for each doubling of the distance in free field. While less indoors, this is a fair rule to use in estimating the separation which can be realized through spacing of musicians.

MICROPHONE MANAGEMENT

This principle of separation by distance applies also to microphones. The placement of the microphone for musician A and the microphone for musician B must be considered along with the actual positions of A and B. In some cases there is a directional effect associated with certain musical instruments which can be used to advantage, certainly the directional properties of microphones can be used to improve separation. The distance between adjacent musicians and the distance between microphones are obvious factors as well as the distance between each musician and his or her own microphone. There is an interplay between all these distance effects and microphone directivity. The nulls of a cardioid or bidirectional microphone pattern may save the day in controlling a troublesome cross talk problem.

BARRIERS FOR SEPARATION

Physical separation of musicians, absorbent studios, and proper selection, placement, and orientation of microphones are limited in the degree of acoustical separation they can produce. Baffles (or screens as they are sometimes called) are used to increase isolation of the sound of one musician from that of another. Baffles come in a great variety of forms: opaque and with windows, reflective and absorbent, large and small. Extreme forms of barriers are nooks and crannies for certain instruments or a separate booth for drums or vocals.

The effectiveness is very low for baffles of any practical size at low frequencies. Once more we come up against the basic fact of physics that an object must be large *in terms of wavelength of the sound* to be an effective obstacle to the sound. At 1 kHz the wavelength of sound is about 1 foot, hence a baffle 6 feet wide and 4 feet tall would be reasonably effective. At 100 Hz, however, the wavelength is about 11 feet and a sound of that frequency would tend to flow around the baffle, irrespective of the thickness or material of the baffle.

ELECTRONIC SEPARATION

Some use, although not extensive, has been made of electronic gating circuits to improve separation between sources. Such circuits reject all signals below an adjustable threshold level.

ELECTRONIC INSTRUMENTS AND SEPARATION

Contact pickups applied to almost any musical instrument with a special adhesive can transform an acoustical instrument to an amplified instrument. In addition, there are many electrical instruments which are completely dependent on the pickup transducer and amplification. The electrical output from such instruments can be fed to the console, providing dependence on the electrical signal rather than microphone pickup of an acoustical signal. The separation between two such tracks can be very high. Amplified instruments having their own loudspeakers in the studio can be picked up by microphones placed close to the loudspeakers. Even though the quality of such sound is degraded, this approach has its enthusiastic followers who like the sound.

THE FUTURE OF MULTICHANNEL

Multichannel recording techniques have production advantages which promise to be a permanent part of the recording scene of the future. Although television is, at the present, strictly monophonic, the audio control room of a network production center might have a mixing console with 50 channels and multitrack recording facilities to match. Recording a symphony orchestra today commonly requires many channels and many tracks. In this case it is not true separation recording, but rather single point recording for ambience with augmentation of certain instruments, sections, and soloists as required. At this time there seems to be a trend away from the artificialities of strict separation recording in some areas.

AUTOMATION

The number of knobs, switches, buttons, and VU meters on the average recording console is enough to dazzle the uninitiated and impress the musically oriented client. In fact, it becomes something of a problem for an operator equipped with only one pair of eyes, one brain, two hands, and a normal complement of reaction times to operate all these controls and still have time for the more creative aspects of his job. Minicomputer automation control relieves the operator of much of the tedium and releases him for more creative work. Automatic computer controlled adjustment of pot settings, equalization, etc., in a mixdown, especially, is a boon to the operator, making him a far more productive worker. Development in the direction of greater automation is inevitable for the future.

An introduction to the literature of multichannel recording may be found in References 110 through 116.

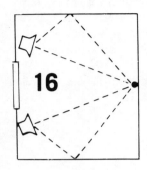

16   The Control Room

There is nothing "ho-hum" about the subject of control rooms. At the present time a very healthy upheaval is taking place in which accepted practices of the last 40 years are being questioned. Quotations from three highly respected leaders in the audio field concerning control room treatment show the present state of the art:

"Provide as much absorption as practical but apply it uniformly to all surfaces."[117]

"The frontal part . . . of the room should display some reflective panels . . . the rear wall . . . should receive an abundance of sound absorption . . . "[118]

"The absorption is to be placed . . . at the front of the control room . . . The remainder . . . to be without absorption."[119]

Absorbing material applied uniformly to all surfaces, largely confined to the rear wall behind the operator, or entirely on the front around the monitors - what will it be? This is not evidence of capriciousness or floundering about or an expression of personal taste. Each of these three men has given much thought to the basic problem of providing the mixing operator with the best possible acoustical environment to enable him to hear the studio sounds unencumbered by acoustical flaws of the control room. If he makes equalization or level adjustments to compensate, unknowingly, for faults in his room, the product is degraded and his mix will not sound right in other environments. Interestingly, there is solid acoustical and psychoacoustical reasoning behind all three points of view.

UNIFORM DISTRIBUTION OF ABSORPTION?

Queen,[117] who suggests applying the absorbing material uniformly on all surfaces, points out the basic differences between a large space and a small one such as the control room. In the large auditorium, first the direct sound reaches the observer and the reflections arrive considerably later because of the great size of the room and remoteness of walls, ceiling, and other reflecting surfaces. In a small control room the reflected components follow close on the heels of the direct sound. This is a very important distinction when the fusion period of the human hearing mechanism is considered. The direct sound and all reflections arriving within about 65 milliseconds are fused into a single, unified perception. These early reflected components do not sound like echoes, but rather are perceived as part and parcel of the direct sound, enhancing its character and loudness. Reflected sound arriving after 80 or 90 milliseconds, if an isolated, discrete reflection, sounds like an echo; if many reflections it is heard as reverberation. All three men concede these basic psychoacoustical phenomena.

Queen then goes on to consider the distortion of the frequency spectrum of the room itself. Room modes affect the level of sound as position is changed, and these peaks and valleys are very much frequency dependent. Thus the spectral balance of the sound changes with position. Even more than this, spectral quality depends on the change of absorption in the room with frequency which, in turn, affects the spectral quality of reflected sound. He then recommends that the absorption in all octave bands be balanced and suggests the possibility of an indoor-outdoor carpet with an open-cell foam pad on all wall and ceiling surfaces. Interested readers should refer to the article[117] for Queen's other recommendations on monitor loudspeakers and procedures for testing the room.

REFLECTIVE CONTROL ROOM FRONT END?

Although quite open to the implications of new findings, Rettinger[118] is not one to be "tossed to and fro with every wind of doctrine". With several decades of studio and control room experience behind him he hesitates to accept the live end - dead end philosophy for control rooms (to be discussed fully in the next section), at least without a thorough consideration of the advantages of a reflective environment for the monitor loudspeakers. He points out that the comb filter interference produced by the direct component and a single reflected component can readily be observed and analyzed. When numerous first order reflections combine with the direct sound at the listener's ears, however, the quantitative and qualitative effects are much more difficult to evaluate. He also believes that the shorter the time intervals between the individual reflections and the direct sound the less the ear is able to detect acoustical comb filter effects. Phasing effects, such as those inherent in the music of many violins in a symphony orchestra, are pleasing to the ear and very much involved in many pleasant everyday listening experiences. Rettinger also seems inclined to be cautious in accepting the results of any one-eared (microphone) measuring system, no matter how sophisticated and he points out that Haas himself conceded that reflections give "liveness" and "body" to small room sound.

Rettinger emphasizes that a highly absorbent environment for the monitor loudspeakers, first, is a partial fiction because of the inevitable presence of the reflective glass surfaces of the observation window and, second, that it is detrimental to the functioning of the loudspeakers. For example, at 100 Hz the directive pattern of the loudspeaker is practically a hemisphere as shown in Fig. 16-1. This means that, in spite of the aiming of the loudspeaker, at low frequencies almost as much sound is directed to the window as to the operator. This reflected low frequency energy arrives at the

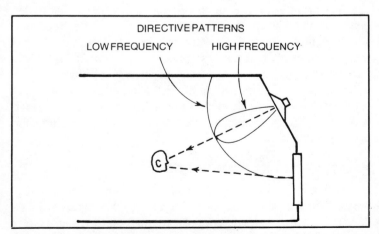

Fig. 16-1. The lack of directionality of control room monitoring loudspeakers at low frequencies results in the mixer hearing low frequency reflections from the glass without associated high frequency components.

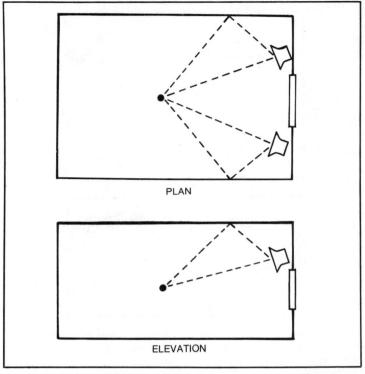

Fig. 16-2. A rectangular control room shape results in excessive delay between direct component and first order reflections.

operator's ears without any accompanying close first-order mid-range or treble reflections. This distorts the signal in a way in which the operator may feel the need to attenuate bass when the signal really does not need it.

REFLECTION MANAGEMENT

Rettinger then expresses his convictions on the need for early reflections in specific suggestions for control room design. Figure 16-2 shows plan and elevation of direct and first order reflections in a conventional rectangular control room and points out that the delay of the reflections is greater than it need be. Reflective surfaces of walls and ceiling must be carefully placed. By the simple expedient of introducing a few new reflective surfaces as shown in Fig. 16-3, the time delay of first order reflections is greatly reduced. This is a principle which can readily be expressed in control room design and construction.

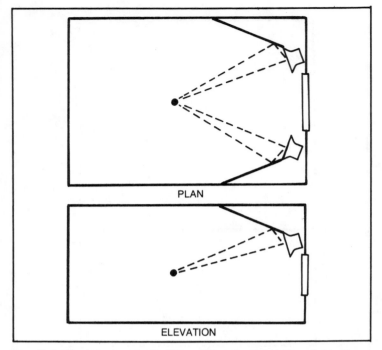

PLAN

ELEVATION

Fig. 16-3. Carefully placed reflective surfaces near the loudspeakers can reduce the delay between the direct component and first order reflections.

RETTINGER CONTROL ROOM DESIGN

Figure 16-4 puts the principles discussed above to work in a control room of about 5000 cu ft volume designed to give minimum delay between direct and early reflected components. This is accomplished by judicious shaping and placement of reflective surfaces above and to each side of the operator's position. Convex shapes increase the floor area covered by preferred reflections so that good listening conditions prevail over appreciable areas, both at the console and at the client-observer seat behind the console.

Reflections of long delay from the rear wall are minimized by making this wall highly absorbent. Wedges of absorbing material used in anechoic chambers could be used except their length (3 ft or more) requires excessive space. Fully satisfactory approximations, much less costly and more compact, suggested by Rettinger, are shown in Fig. 16-5. Two of these built around 2 x 8 or 2 x 10 inch studs are highly effective, the one utilizing 2 x 6 inch studs is somewhat less effective, but satisfactory for many installations. The absorbing material could be Owens-Corning Type 723

Fiberglas of 3 lbs/cu ft density or its equivalent in 2 inch and 1 inch thickness. The entire rear wall is covered with this highly absorbent, wide band absorber. A cosmetic cover of loudspeaker grille cloth adds class to the wall. On the lower 6 ft of the wall subjected to abuse a perforated metal sheet or expanded metal lath should be mounted behind the cloth cover. The airspaces improve low frequency absorption.

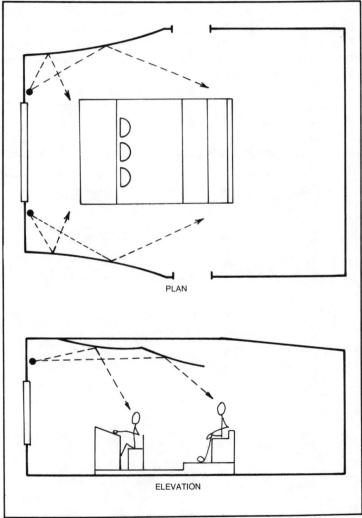

PLAN

ELEVATION

Fig. 16-4. Control room design suggested by Rettinger which provides desirable early reflections over a considerable area (after Rettinger, Reference 118).

LIVE END - DEAD END CONCEPT

In Britain and other countries radio drama is still alive, long after its death in the United States. From the very early days live end-dead end studios have been used to provide the variety of acoustical environmental effects required to enhance dramatic realism. Gilford[34] describes a drama studio complex having a very live room, a very dead room, and a normal talks studio for narration which are supplemented by a long, narrow studio divided into three sections, one live, one medium, and one dead. Double curtains of heavy sailcloth separated about 3 feet are mounted at the two transition points and they are capable of being extended or retracted. When retracted a single long studio with "tapered" acoustics results, with a live end and a dead end.

This descriptive phrase (live end - dead end) is taking on new meaning, principally as a result of observations made by time delay spectrometry. Don Davis[119] feels that absorptive rear walls in control rooms is a direct hangover from the practice of 40 years of making the rear walls of large auditoriums absorptive. He points

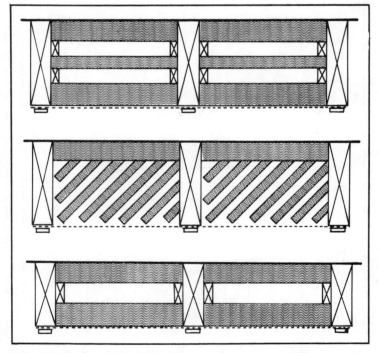

Fig. 16-5. Possible absorptive wall designs for the rear control room wall in Rettinger's control room plan (after Rettinger, Reference 118).

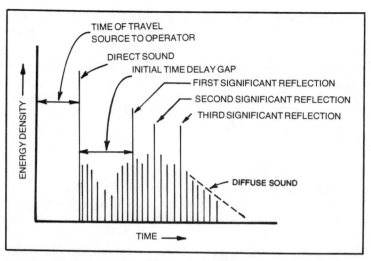

Fig. 16-6. Reflection pattern resulting from a live end - dead end control room design (after Davis, Reference 122).

out that justification of this practice in large spaces does not apply in the much smaller control rooms because of response anomolies resulting from too many early reflections and too few late ones. Davis is inclined to consider studio treatment "a relatively free art form" directed toward any "new sound" which sells records or pleases musicians but that the control room is a space in which accuracy is desired. At least, this is true ideally even though multitrack practices may have blurred the goal of accuracy in recent years.

Observations by Davis in many control rooms by the technique of time delay spectrometry have convinced him that most control rooms are so dead acoustically that no reverberant sound field can be developed, there are too many reflective surfaces near the loudspeaker, the rear walls are too absorbent, and adequate diffusion is absent.

As for dead acoustics, in his study Davis found reverberation times of control rooms ranging from 0.13 to 0.20 second. He feels that a semi-reverberan condition should exist, something between free field (6 dB sound pressure level drop for each doubling of the distance) and a fully reverberant field in which there is no change in level beyond the critical distance at which the direct and reverberant fields are equal. For a 4000 cu ft studio with ceiling height of 10 ft he gives an optimum reverberation time of 0.47 second.

As mentioned earlier, he advises placement of all the absorbing material in the front portion of the control room, roughly from the operator forward and applying it to all surfaces, ceiling, walls, and floor. All surfaces of the rear part of the control room are to be reflective and diffusive. Surfaces should, as commonly agreed, splay symmetrically from front to rear. This arrangement of absorption, reflection, and diffusion is all designed to accomplish a certain time pattern of reflections at the operator's position.

INITIAL TIME DELAY GAP

The initial time delay gap idea applied by Beranek[53] to music halls is applied by Davis[122] to control rooms. This gap is defined as the time gap between the arrival of the direct sound at the operator's ears and the critically important first significant reflections as illustrated in Fig. 16-6. Of course, the initial time delay gap is greater in music halls than in smaller control rooms. In fact, it is greater in most studios than control rooms and Davis deplores this. He feels that the initial time delay gap of the studio should not be masked by that of the control room. To achieve this, his goal is to make the gap in the control room, as illustrated in Fig. 16-7, longer than that of the studio. If such conditions prevail, the operator clearly hears what comes from the studio, unmasked by control room effects. The whole point of the soft front end and reflective/diffusive rear is to lengthen the initial time delay gap of the control room so that the gap of the studio may be heard.

THE DAVIS CONTROL ROOM

We have seen that the basic live end - dead end control room includes a very absorbent front end and a reflective and diffusive area behind the operator. This diffusion is obtained by geometrical irregularities of the rear surfaces. These are commonly of wood or plywood surfaces arranged in 3 dimensional triangular or semicylindrical cross sections. Beside these fundamental aspects, Davis suggests other desirable features such as:

—Positioning the operator 10 to 12 feet from the two monitor loudspeakers which are 12 to 14 feet apart.

—Arranging the operator's head to be about 10 feet from the rear wall, rear ceiling, and rear side walls. This results in a difference in time of arrival between the direct sound and the first diffuse reflections from the rear of about 20 milliseconds, which is within the Haas fusion zone.

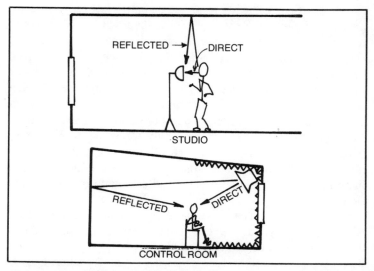

Fig. 16-7. The wider time delay gap of a live end - dead end control room avoids masking that of the studio (after Davis, Reference 122).

He also encourages the use of the time aligned loudspeakers in the control room and pressure zone microphones in the studio to further clean up sound problems.[31]

VEALE'S CONTROL ROOM

As a result of much serious thought and investigation of control room acoustics as related to the human hearing mechanism, Veale, an acoustical consultant in England, has advanced his ideas of what constitutes a good control room.[77] In Chapter 10 his conviction was mentioned that only inert, dissipative type of sound absorption should be used in a control room. This would rule out Helmholtz absorbers and resonant bass traps of all kinds and, on the surface, would seem also to require solid concrete and masonry construction rather than frame construction which absorbs significantly in the low frequency region by diaphragmatic action.

Veale has specific ideas on the ideal reflection pattern at the operator's position. He feels that reflections arriving at the ear within 8 milliseconds of the direct sound serve no useful purpose in this context. Reflections arriving later than 10 milliseconds are useful in creating the sound picture. With regard to music, sounds arriving later than 80 milliseconds create the room ambience. Therefore he states that reflections arriving at the ear between 10 and 70 milliseconds are useful in a control room.

In practice, Veale has observed that if too few reflections are provided in the control room, the resultant product (when listened to under domestic conditions) is often lacking in reverberation content. If too many reflections are provided, the reverberation on the product is either excessive or of wrong texture. He concludes that between 4 and 7 reflections are required by the ear to give a complete sound picture. The first reflection should arrive between 10 and 15 milliseconds after and be 4 to 6 dB lower than the direct. Remaining reflections should be evenly spaced out to 50-70 milliseconds and their peak levels should follow a 350 dB/second decay rate (reverberation time of 0.17 second). He then describes instrumentation to measure delay and amplitude of such reflections.

Unfortunately, Veale gives no specifications for a typical control room which yields the reflection pattern suggested. He does, however, discuss design factors in general. He requires that the front portion of the control room between the loudspeakers and the operator generate the primary reflections and, "unless sufficient reflections exist, to treat the remainder of the room for total absorption". He considers the ceiling to be important in establishing the desired reflection pattern. It would appear that Veale generally agrees with Rettinger as to the importance of early reflections. The reflections between 0 and 8 milliseconds, which Veale characterizes as having no useful purpose, are, in fact, those comb filter producers which Davis seeks to minimize by an absorptive front end.

EXPERIENCE WITH LIVE END - DEAD END CONTROL ROOMS

Several control rooms have been constructed along the live end—dead end principles expounded by Davis and with his direct involvement. Those associated with these projects report "a new realism that you can only experience from live performances".[120] "(It) takes a little getting used to. There is an open, transparent feeling as though you are . . . in the studio with the musicians".[121] Good stereo imaging over a wide area is reported as well as minimum change in spectral balance anywhere in the rear part of the room. Problems of microphone selection and placement, normally masked by control room problems, are said to become glaringly evident. "It is a strictly clinical atmosphere; there's no hype".[121]

COMPARISON OF VIEWS

It is very interesting that the divergent views of anechoic rear and anechoic front control room areas stem from divergent in-

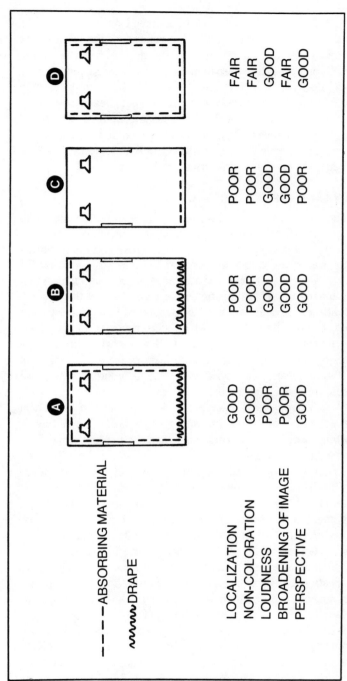

Fig. 16-8. Summary of results of Nippon Gakki psychoacoustical experiments. The judgements of four audio professionals are listed for four arrangements of acoustical absorbing material (after Kishinaga et al, Reference 97).

terpretations of the effect of early reflections. Don Davis says, "the shorter the delay experienced by the reflection, the broader the individual response anomolies constituting the comb filter observed"[31] and that these broad anomolies are more audible than narrow ones. Michael Rettinger[123, 124] would seem to agree with this as far as a single reflection is concerned but goes on to express his view that many short delay reflections combined with direct sound and heard binaurally is what makes good music sound good. It is evident that what is needed to settle this point are carefully designed and controlled psychoacoustical tests.

NIPPON GAKKI EXPERIMENTS AGAIN

The psychoacoustical experiments[97] referred to in Chapter 13 in the context of the home listening room may also shed some light no the present control room discussion. Kishinaga and his associates report the judgements of four audio professionals on the quality of a variety of musical recordings as they listened in test rooms having reverberation times between 0.3 and 0.4 second. Eight different arrangements of the acoustical environment were studied, the results of four of which were reported in his paper. These four are described in Fig. 16-8 along with the average judgements of the observers. Some liberties have been taken in Fig. 16-8 in substituting "good", "fair", and "poor" for the more scientific form Kishinaga used.

Wide extremes of amount and arrangement of acoustical material on the walls are shown in Fig. 16-8. The ceiling of "absorbing board made of rock wool" and the parquet wood floor are constants during the tests. Which arrangement is best depends on the relative importance one places on the five areas of judgement - localization, non-coloration, loudness, broadening of image, and perspective. Localization of the sound image was best in arrangement A in which sound absorbing material was on three walls and drapery on the fourth. Arrangement A also resulted in the lowest coloration, although Kishinaga reported colorations in all tests. Loudness seems a less important dimension in regard to room treatment as it can be compensated easily by increasing gain. Broadening of image would be a factor more valuable in enjoyment of music than in the critical atmosphere of the control room. Early side wall reflections of B and C yielded good broadening of image. In summary, the experiments of Kishinaga et al seem to indicate that sound image localization, relative freedom from coloration, and perspective offered by treatment A in Fig. 16-8 are superior to the other treatments. This may be considered at least partial support for Don Davis' absorptive front end concept for the control room although it lacks confirmation of the diffusive rear end. It is interesting that absorptive side walls seem to be a more important factor than whether the wall behind the loudspeaker (B) or the rear wall alone (C) are absorptive, yet D with all walls absorptive (except the one behind the loudspeakers) yields poorer results than A.

BUDGET LIVE END - DEAD END

Is it possible to apply the live end - dead end principle to budget control rooms and reap at least some of the promised values, even though numerous compromises are involved? Such a control room is pictured in Fig. 16-9. The walls and ceiling are rectangular and not splayed, the room is smaller than desired, but the client is happy and tests indicate, perhaps, at least some of the satisfaction results from the fact that a reasonable initial time delay gap has been achieved.

Figure 16-10 is an echogram taken in the control room of Fig. 16-9. The omnidirectional measuring microphone was placed at operator ear position. An air pistol impulse source which ruptures a paper diaphragm was positioned at the face of the left monitor loudspeaker. The direct sound is the large negative going pulse at time equals 0. At 15 milliseconds after this a large reflection only

Fig. 16-9. A budget live end - dead end control room. Walls and ceiling from console forward are covered with 4 inch thickness of dense glass fiber. The rear of the room is reflective with triangular plywood diffusing elements on rear wall (World Vision International).

about 2.9 dB down from the direct pulse arrives, followed by a host of others 6 to 8 dB down in the time interval from 15 to 25 milliseconds. But during the interval from about 3 to 15 milliseconds there is a definite reduction in amplitude of reflections and this is the initial time delay gap of this room.

Peak amplitudes of reflections of Fig. 16-10 are plotted against time in Fig. 16-11. Except for the first 15 milliseconds the reflected pulses follow the general slope representing the reverberation time of the room, 0.35 second. There is a definite gap during the first 15 milliseconds which can, hopefully, be considered a useful initial time delay gap achieved on a budget basis.

ELECTRO-ACOUSTICALLY COUPLED SPACES

We have considered acoustically coupled spaces in the home listening environment (Chapter 13) and in auditoriums (Chapter 8). The coupling instrument in these cases may be something as simple as an open doorway. The control room and studio are also coupled acoustically, but in this case the coupling is electrical by way of the microphone/amplifier/loudspeaker path as shown in Fig. 16-12. Nevertheless, a true coupling exists, even though it is a one way affair, studio to control room. What effect does studio reverberation have on control room reverberation through this electrical link?

Mankovsky[69] discusses the question. When the control room is connected electroacoustically to the studio, the effective reverberation time of the control room is increased over that of the control room alone. Interestingly, this combined decay is not exponential. The resultant control room reverberation time is dominated by that room having the longer reverberation time, but,

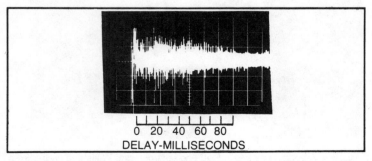

Fig. 16-10. Echogram taken in control room of Fig. 16-9. Impulse source at face of left loudspeaker, omnidirectional measuring microphone at operator's ear position.

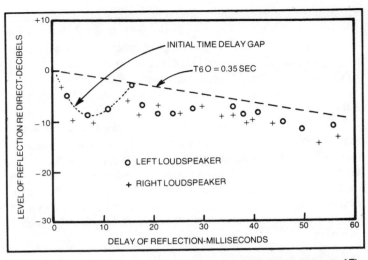

Fig. 16-11. A plot of level vs. delay of the peak reflections of the echogram of Fig. 16-10 showing a 15 millisecond initial time delay gap resulting from the live end - dead end treatment of the control room of Fig. 16-9.

because studio and control room reverberation time are usually reasonably similar, the magnitude of this effect is small. Let us consider the case of film sound recorded in a small enclosure played back in a large theater. The longer reverberation time of the theater would dominate the combination and be slightly longer, and slightly less exponential.

CONTROL ROOM VOLUME AND SHAPE

Good distribution of normal modes should be a basic rule in any listening room, especially in control rooms. This calls for

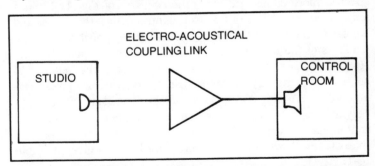

Fig. 16-12. Diagram of the electroacoustical coupling between studio and control room. Studio reverberation time alters control room reverberation time via this one way link but the effect is nominal.

reasonable volume, but there are no laws stating that the volume should be thus and so. Volumes in the 2000 to 5000 cu ft region undoubtedly include a high percentage of control rooms of recording studios today. A consideration of the desired reflection pattern (Rettinger type, Davis type, other) bears on the distance of side and rear walls from the console position and thus will affect the size of the room.

The reflection pattern desired determines the shape of the room. Even though a basic rectangular shape is available initially, the construction of walls and ceiling reflective surfaces within that rectangular space provides the surfaces to achieve the pattern. A trapezoidal shape, narrower at the window end and widening toward the rear is almost universally used. A drop ceiling of carefully designed angular surfaces can provide a major share of useful reflections.

CONTROL ROOM NOISE SPECIFICATIONS

Control rooms are cursed with the noise of equipment, principally that of the all important recorders. To that must be added air conditioning noise, considered in detail in Chapter 12. Other people in the control room produce noise, as well. Even though some noise contributors are necessary and unavoidable, something usually can be done to reduce most background noises which serve only to divert attention and degrade the judgements of the mixer-operator.

The noise criteria curves of Fig. 12-1 should be helpful in establishing a noise specification for control rooms. A few precautions are in order in their use, however. Often tonal components rear their ugly heads above the noises which have distributed spectra, more or less following the contours of Fig. 12-1. Such single frequency hums or buzzes can intrude upon the senses and we need a method of measuring their relative contribution to the loudness of the overall noise. The A-weighted response of Fig. 16-13 is designed to give us a rough indication of the loudness of noise as perceived by the human ear. Sound pressure level readings with the A-weighted network switched in will give us at least a very approximate evaluation of loudness of a noise with a prominent single frequency component. The relation between the NC contours and A-weighted sound pressure levels is given in Table 16-1.

Octave analysis of the background noise, required to apply Fig. 12-1, often fails to reveal the presence of prominent single

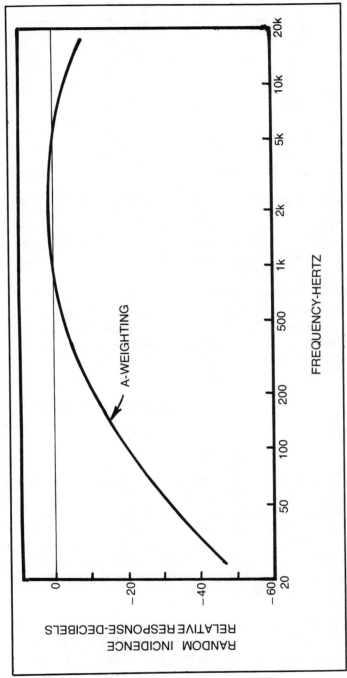

Fig. 16-13. The response of the A-weighting network of sound level meters. The sound pressure level (A-weighting) reading conforms very roughly to perceived loudness.

Table 16-1. A-Weighted SPL and NC Contour Relationship.

Noise Criterion Contour	Sound Pressure Level (SPL)-dB (A-Weighting)
NC-20	30
NC-25	35
NC-30	40
NC-35	45
NC-40	50
NC-45	55
NC-50	60
NC-55	65
NC-60	70

frequency components. First plot octave sound pressure levels on Fig. 12-1, then take an overall sound pressure level with the A-weighting network, and compare the two through Table 16-1. The presence of tonal components should be suspected if NC contour and SPL (A-weighted) do not agree.

CONTROL ROOM REVERBERATION TIME

The approach to control room treatment (absorptive front, absorptive rear, etc.) will affect the reverberation time goal. As stated previously, Davis found control room reverberation times from 0.13 to 0.20 in his investigation, yet he suggests 0.47 second in one example. Rettinger[118] suggests a reverberation time of:

$$RT60 = 0.15 \log V - 0.15 \qquad (16\text{-}1)$$

where, RT60 = reverberation time, seconds

V = volume, cu ft

A control room of 2000 cu ft, by this rule, should have a reverberation time of 0.35 second, one of 4000 cu ft 0.39 second. Fierstein's opinion on the subject is summarized in Fig. 16-14.[125]

Whatever the basic 500 Hz reverberation time, and whatever philosophy of control room design followed, all agree that reverberation time is an important factor and that it should be essentially uniform with frequency so that all components of the signal die away at the same rate.

MONITOR LOUDSPEAKERS

Loudspeakers are commonly dubbed the weak link in the recording and reproducing chain; others would reserve this ques-

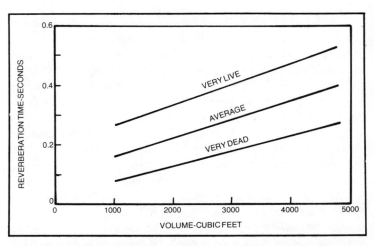

Fig. 16-14. Fierstein's opinion on the range of acceptable control room reverberation time (after Fierstein, Reference 125).

tionable distinction for room acoustics. Both are important and both can cause problems if not right. High quality loudspeakers are imperative for critical monitoring. It turns out that time and phase accuracy in monitor loudspeakers is of utmost importance.

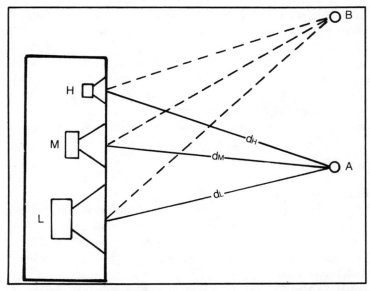

Fig. 16-15. In a properly aligned monitor loudspeaker the acoustical outputs of the high, midband, and low frequency units will arrive at the observer's ears (at A) at the same time.

A glance at Fig. 16-15 will help us appreciate this fact. Here is a schematic of a three-way loudspeaker composed of high frequency, midrange, and bass transducers. The signal spectrum is broken down into these three sections, each driving its own transducer. A listener at A receives these three slices of the audible spectrum over different paths and if the three path lengths differ there is distortion of the combined signal. A listener at B would experience quite a different distortion due to different times of arrival. However, in control rooms we are primarily interested in the symmetrical arrangement as the monitor loudspeakers are carefully aimed at the operator at the console.

Figure 16-15 is hopelessly inaccurate because the source of the sound is not out at the grille cloth as indicated, it is somewhere back in the bowels of the H, M, and L transducers, but where? At one time the voice coil was considered the point of origin of the sound of each unit and they were aligned accordingly, but measurements with sophisticated equipment proved this not to be the case. Nor are we so much interested in just where this point is, rather we are very interested in adjusting the sound from all three transducers to arrive at point A at the same time. The problem is that the acoustical pressure wave does not emerge from the loudspeaker immediately upon electrical excitation, but after a certain delay. This delay is dependent on frequency in a complicated way and, if not so corrected, results in a time smearing of the acoustical image.

Crossover networks, which slice the spectrum into pieces appropriate for efficient radiation by the high, midband, and bass transducers, can be designed to introduce just the right delay to align the signals for simultaneous arrival at some fixed point A. The improved fidelity of sound reproduced over such an aligned system is a step forward in control room sound quality. [126]

LOUDSPEAKER MOUNTING

The method of mounting monitor loudspeakers affects their reproduced frequency response. Hanging loudspeaker cabinets from the ceiling or mounting them on wall brackets affects the low frequency response of the units adversely due to cavity resonances between and behind the cabinets. A far better way is to build the loudspeakers into carefully shaped inclined planes with faces flush with the surface.

EQUALIZATION

The acoustical response of a control room (or any other room) is a valuable factor. It includes the acoustics of the control room and the loudspeaker's action in that particular room, as contrasted to an anechoic chamber in which the manufacturer obtained his nice flat curves that sold the loudspeaker. The elements of the acoustical response procedure are given in Fig. 16-16. Pink noise drives the power amplifier and the loudspeaker energizes the room. The microphone of the measuring system is placed at the operator's ear position.

A laborious but inexpensive approach is to use bands of pink noise (usually ⅓ octave) and measure the level at the operator's position with a sound level meter. If the sound level meter happens to have filters built into it, wideband pink noise can be used to excite the room with the frequency analysis taking place on the receiving end. The pink noise can come from a noise generator or from recordings made personally or purchased from commercial sources.[99]

A far more convenient approach to acoustical response in a control room is the use of a real time analyzer (RTA).[127, 128, 129] The pink noise in the room is wideband and filters within the analyzer reveal the acoustical response of the system in each band by a set of bars on a cathode ray oscilloscope or on vertical columns of light emitting diodes (LEDs). The real time analyzer exhibits a plot of room response. It is common practice to use a multiband boost/dip equalizer in the recording circuit to compensate for response irregularities revealed by the real time analyzer or the point by point method.

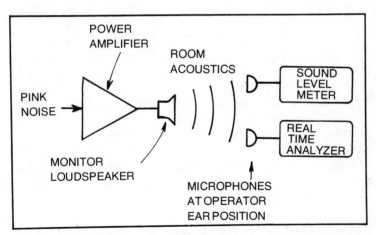

Fig. 16-16. Arrangement for measuring the acoustical response of a room. The value of real time analysis in control rooms is limited. Equalization cannot correct basic acoustical flaws of the room.

311

To become oriented properly on the application of the real time analyzer system to control rooms we should realize that they were first used in sound reinforcement. In an auditorium, system howling at one frequency can be stopped by a dip filter which reduces gain at that frequency. Increasing system gain puts the system into feedback at some other frequency which can then be stopped by another dip filter at the new frequency. When the process is completed considerable improvement is realized in system gain before feedback with only a few narrow notches removed from the spectrum. Feedback is not a problem in control rooms, but equalization was applied to adjust control room acoustical response to the desired shape. There are proponents and opponents to such procedures. One group declares that properly treating the control room and using high quality monitor loudspeakers eliminates the need for equalization, or at least very much of it. [130] The other group says equalize. In striving for the full 20 Hz - 20 kHz band, problems often appear at the extremes and here the complications of equalization also appear. To correct for a low frequency roll off, a doubling of power is required for each 3 dB boost at the console. The high frequency tweeter power handling ability may be a limiting factor on that end.

EQUALIZATION LIMITATIONS

Perhaps the most telling argument against the use of any very great amount of equalization in control rooms has to do with the very nature of the acoustical response measurement, whether point by point or real time analyzer. This is strictly a steady state measurement and many acoustical problems of control rooms are transient in nature. Sound decay rate problems cannot be solved by flattening a steady state response curve by equalization. [131] The desirability of uniform reverberation throughout the band has been emphasized. If a big rise in low frequency reverberation time exists, reducing low frequency response to cure "boominess" is an exercise in futility. The wise conclusion to all this seems to be that equalization may be helpful to correct minor loudspeaker humps or hollows revealed by anechoic measurements, but that defects due to loudspeaker deficiencies or mounting or room acoustics had best be treated at the source.

CONTROL ROOM STANDARDIZATION

It would be very desirable to be able to take a tape mixed in one control room and have it sound the same when played in another control room. It can be said that all our striving for better control room conditions is toward (a) more accurate listening conditions which, it is hoped, will yield (b) more uniform conditions from one control room to another. Item b is taking on increasing importance as few recording operations are isolated islands of activity.

The classic case of sound mixing standardization is that of the Academy of Motion Picture Arts and Sciences which, in the early days of sound films, established the "Academy curve", a frequency response curve for film sound reproduction in theaters. This did succeed in holding things together, but critics contend that improvements have been stifled in the process. The Nordic Film and Television Union, for example, active for a decade, has a draft of an international standard covering the electroacoustical response of motion picture control rooms and indoor theaters.[132] These standards are not directly applicable to the usual control room. It makes a great difference whether the final product is heard in a large theater or in a small home living room. While it is interesting to follow such attempts at standardization, there is little effort being expended in this direction for control rooms for studios dedicated to television program and recorded music production. True interchangeability of product, however, depends upon such eventual standardization.

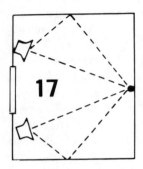

Smart Instruments For Acoustical Measurements

17

I often say that when you measure what you are speaking about, and can express it in numbers, you know something about it; but when you cannot express it in numbers, your knowledge is of a meager and unsatisfactory kind. It may be the beginning of knowledge but you have scarcely, in your thought, advanced to the stage of Science whatever the matter may be.

Lord Kelvin

A scientific field advances at a speed determined by the availability of adequate measuring instruments. "To measure is to know." These truisms apply to the audio field as well as others. Digital techniques are fast revolutionizing recording processes and they are in the process of preempting the test instrument field. Digital techniques are now making their presence felt in the field of measuring instruments through the application of the microprocessor and it is well for us to be informed about what is happening.

THE MICROPROCESSOR AND MICROCOMPUTER

The statement has been made that computers can do nothing for us that we cannot do ourselves. Right and wrong. That is like saying traveling by airplane is like walking. True, both forms of transportation are capable of getting us from here to there, but some trips are practical only by flying. The element of speed is very important in this accelerated age, and this surely must apply to acoustical measurements in the years ahead. Speed is not the only factor. There are some measurements than can only be made with advanced technology.

The microprocessor is an information processing unit which executes user instructions. With random access memory (RAM), read only memory (ROM), and interfacing input/output (I/O) devices we have a system called a microcomputer. This is the hardware. Programming the microcomputer to do a particular job is done with software. Microprocessors, now available on a single chip, are presently applied in a host of trivial fields such as video games and pinball machines and to more useful fields such as TV tuners, clothes washers, and lawn sprinkler timers. Microcomputers are also being incorporated in acoustical measuring equipment.

The following sections describe six different advanced measuring systems employing digital techniques. Some are hybrid devices, a mixture of analog and digital. Four of them are on the market as this book goes to press; the marketing plans for the Sony and Yamaha systems to be described are not known. A stream of microprocessor-controlled measuring instruments is now appearing on the market. The six to be described are illustrative of the new level of measurement convenience and speed soon to be commonplace.

Heyser's time delay spectrometry[135] and numerous other allied systems, some involving the fast Fourier transform (FFT), are on the horizon. They represent the trend toward more sophistication in conception, greater analyzing power, greater portability, and convenience. It will be interesting to see if all these tremendous positives can be incorporated units which can be marketed at a reasonable price.

AMBER MODEL 4400-A

Starting alphabetically, this instrument is manufactured by Amber Electro Design, Ltd., of Montreal, Canada H4P 2N5. The Amber Model 4400-A multipurpose test set (Fig. 17-1) contains virtually an electronics and acoustical laboratory except X-Y recorder and/or cathode ray oscilloscope for plotting data. All this is one 30 pound, 7" x 16" x 16.8" package. It has been designed to meet a growing demand for faster, more convenient and more comprehensive measurements to match the increased sophistication of modern audio equipment and the rising interest in room acoustics. It is state of the art in conception and design and specifications would indicate that there are few compromises in performance.

Generator Section

The Amber 4400-A generates a bevy of test signals including sine, triangle, square, tone burst, and pink noise. The sine and triangle may be assymetrically clipped for system polarity verification. Ten crystal controlled sine waves at octave spacings are available for single or combined use for markers on oscilloscope readout. Flexible zero crossing gating circuits can form pulses of 11 gate times and pulses may be repetitive. A logarithmic sweep of frequency over the ranges 20 Hz to 20 kHz or 100 Hz to 100 kHz is provided. Time base sweeps of from 1 to 1024 seconds (17 minutes) are available. A unique feature is the low distortion of the sine waveform, less than 0.05% in the frequency range 100 Hz to 10 kHz. A crystal stable comb signal consisting of any combination of 10 sine waves at octave intervals from 31.5 Hz to 16 kHz is available, among other things, for tape speed tests. In addition, a high power output stage is included (100 volt peak to peak swing, +33 dBm) for system headroom tests.

Meter Section

The meter section of the 4400-A provides digital measurement of frequency and level of both the generator output and the receiver input with commendable accuracy. Frequency may be measured from 10 Hz to over 150 kHz. The autoranging digital level meter has a range from over +30 dBm to − 120 dBm (narrow band) or − 90 dBm (wideband). Measurement resolution is to 0.01 dB. True RMS as well as peak or average detection is incorporated. The following weighting networks are available: 20 kHz low pass, optional high or low pass filter, ANSI A, B, C, and an internal accessory socket may be included as an option for other user defined filters.

Receiver Section

Plots of amplitude or phase vs. time or frequency may be stored in up to four storage memories and displayed on any standard non-storage oscilloscope. Plots may be generated of wideband amplitude, amplitude difference between two signals, phase shift between send and receive or between two external signals. The vertical axis is linear in dB in 10 dB increments over a 150 dB range. The vertical axis is also linear in degrees over a range ± 60° to ± 180° in 60″ increments.

316

Analyzers

The instrument contains a multimode filter and spectrum analyzer. The filter may be configured as bandpass, band reject, high pass, or low pass and has variable percentage bandwidth. As a spectrum analyzer the band pass filter is logarithmically sweepable and adjustable in bandwidth from approximately 3% to 70%; the 70% corresponds to approximately one octave bandwidth and the 3% to about 1/25th octave. A bandwidth of ⅓ octave corresponds to about 24% of the center frequency. The filter and spectrum analyzer may be used with the digital meter for noise and other types of measurements and with the digital plot recorder for generation of noise floor plots, crosstalk measurements, acoustical response plots, reverberation time plots and others. Acoustical measurements are well served by the logarithmic x-axis and dB y-axis, the variable bandwidth feature, digital programming, and storage.

Swept frequency spectral analysis in the audio band is limited to relatively low speeds by the physics of the situation. Readout of such information normally requires a storage oscilloscope. The four storage memories of the 4400-A may be used to accumulate the information which may then be displayed on a conventional oscilloscope. Level vs. time plots of reverberation decays may also be stored and read out as well. Any two stored records may be read out simultaneously for comparison.

The 4400-A does not have a real time analyzer mode. Its makers point out that a ⅓ octave real time display is relatively coarse and could miss important details which the 4400-A spectrum analysis would reveal.

X-Y Recorder

The Model 441 X-Y Recorder Interface is required to generate hard copy plots on standard X-Y recorders. This device provides the data buffering, logic commands, timing, and x ramp generation to both the 4400-A and the X-Y recorder to permit simple, one button operation. The 441 mounts inside the 4400-A on the rear panel.

BADAP 1

The Badap 1, pictured in Fig. 17-2, is billed as an audio microcomputer for tests and measurements. It was developed by Barclay Analytical but is being offered by Crown, Elkhart, Indiana 46517 in the repackaged BDP-2 form weighing 35 pounds. It is a

Fig. 17-1. The Amber Model 4400-A multipurpose test set.

computer-based system specifically designed to perform audio measurements. It is capable of doing nothing by itself, all functions are controlled by ultraviolet erasable programmable read only memories (EPROMS) which simplify program changes.

Badap 1 is made up of only a few basic sections. First, there is the 9″ color monitor and husky power supply. Then comes the computer, complete with central processing unit and memory. There is also a high speed video generator capable of creating color graphics and alpha numerics. The remaining section contains hybrid analog-to-digital interface circuitry as well as microphone and line preamplifiers. Instead of digital filtering, Badap 1 uses conventional analog filters and analog detectors for reasons of economy. They are working on a digital system and when it is economically practical a simple interchange of printed circuit boards could later change from hybrid to all digital.

The front panel of Badap 1 has 12 push buttons which are labelled only with numbers. The functions of these are defined by the particular program in use through video screen labels. Five fixed function touch controls on the panel are used for POWER, BEGIN (takes program back to start and clears all data memories),

318

PREVIOUS (steps back frame by frame to earlier display or control modes), GRATICULE (generates amplitude divisions calibrated in dB), and LABELS (labels 0 dB point, input sensitivity, ⅓ octave centers, and screen range for the real time analyzer program, for example).

Color Display

The color display makes possible showing more than one color-differentiated display simultaneously. Blue, red, green, and yellow are provided plus four other colors for special situations. A bar graph display (4 colors) may also have small colored dots (4 colors) making a total of 8 possible displays on the screen at one time.

Instead of the detector outputs feeding the screen directly, they are sampled and stored in digital form. This makes possible an accumulate mode by which, for example, the maximum level of a signal may be displayed and it will remain in the memory until erased or the Badap is turned off.

Average and peak displays are available. There is also an accumulate mode for average and peak. A special long term average (called FLOAT) is available for both average and peak.

Memory

The screen display is generated by special circuitry that reads directly from the memory. The memory is divided into 8 sections, one for each color and dot/bar shape. Once the display and type of measurement are selected, the computer feeds data into the appropriate memory and updating takes place. With this system the display may be frozen, but the updating of the memory will continue.

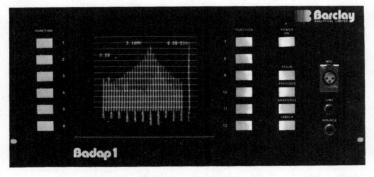

Fig. 17-2. The Badap 1 audio microcomputer for tests and measurements.

Test Signal

Badap 1 does not use pink noise. It uses what is called the DGFS (digitally generated flat spectral) signal. This signal has a crest factor (peak/average ratio) which is constant over the audible band. The crest factor of pink noise is greater at low frequencies than in the high frequency region. The crest factor of DGFS is also several times that of pink noise which stresses the unit being tested more like music does.

Reverberation

The above discussion is slanted toward the real time analyzer program. Installing PROMS containing the RT60 program converts the Badap 1 to measuring reverberation time. The source output is connected to an external power amplifier which drives the loudspeaker. The desired ⅓ octave is selected. The amplifier gain is advanced until a reference cursor on the display blinks signaling the proper level. Touching the RUN button interrupts the signal in the room and the graph of the decay appears on the screen. The computer will calculate RT60 if asked to do so. This smart computer decides how much of the decay is far enough above the background noise to be usable and this is the amount used in the calculations. There are 8 memories in the RT60 program and this means that up to 8 decays may be stored and recalled singly or together.

New Programs

It is the policy of the company to make new programs available from time to time. The real time analyzer program is supplied with Badap 1. The RT60 program was made available later. Many other programs are being considered, including 1/6 octave filtering, various distortion measuring systems, digital filtering to replace the current analog filtering, fast Fourier transform (FFT), loudspeaker alignment, and cassette storage and digital plotting programs. There is substantial programming space in Badap and giving up an existing one to make room for a new one is rarely required. Memory expansion modules are also available. Reference No. 136 describes the Badap 1 in greater detail.

INOVONICS MODEL 500

Inovonics, Inc., of Campbell, California 95008, offer their Model 500 Acoustical Analyzer pictured in Fig. 17-3. This is

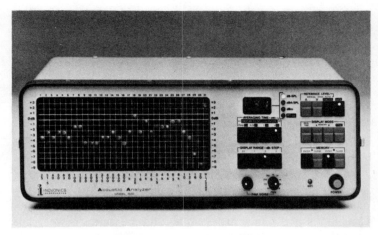

Fig. 17-3. The Inovonics Model 500 acoustical analyzer.

basically a real time analyzer and reverberation analyzer which has
many peripheral uses in noise studies, product development,
equipment testing, adjustment and calibration. The Model 500
utilizes a 13 x 31 LED matrix display and a digital readout for sound
pressure level readings for flat or A-weighted response and reverberation time up to 10 seconds. An external oscilloscope output
is available with vertical deflection signals and sweep trigger. This
gives a vertical bar display.

The analyzer features thirty 2-pole ⅓ octave filters with
center frequencies from 25 Hz to 20 kHz. An internal pink noise
source is available to drive an external amplifier and loudspeaker
for real time analysis. It also provides octave bands of noise from
63 Hz to 8 kHz for reverberation measurements. Octave filters in
the RT60 receive line are slaved to the pink noise filters for
ambient noise rejection. This means that the octave reverberation
decays are displayed on a screen having ⅓ octave resolution.
Reverberation time is also computed and displayed on the digital
readout with 10 millisecond resolution. The value of the LED
matrix display of reverberation decays is inspection of decay shape
to help in judging the trustworthiness of the digital computed
values of RT60.

As an option Inovonics offers an X-Y recorder interface for the
Model 500 so that hard copy records can be made of real time and
reverberation analyses. Digital data from the Model 500 is con-
verted to step function analog X and Y outputs to feed any plotter
with 2 volt full scale sensitivity. The assembly mounts within the

protected recess on the back panel. The PLOT button is on this assembly.

The Model 500 operates from the power mains or from internal batteries. A typical operating life of batteries is 3 hours with an 8 hour recharging time. A review of this instrument may be found in Reference 137 and an operational review in Reference 138.

IVIE IE-30A

Ivie Electronics Incorporated of Orem, Utah 84057, offers a very compact, handheld ⅓ octave spectrum analyzer and precision sound level meter. Their Model IE-30A pictured with the companion IE-17A, to be described later is shown in Fig. 17-4. The functions included in such a small space is amazing and a study of its capabilities gives a new appreciation of the measuring power made possible by the minicomputer.

The display is a 16 x 30 LED array whose intensity is adjusted automatically for ambient brightness. Even the control panel lights up in low light environments.

The omnidirectional electret microphone extends from the case on a 7-inch wand to minimize the effect of sound reflections from the case. The microphone and its internal preamplifier can be used remotely up to several hundred feed using extension cables.

A four digit readout shows sound pressure level in dB with A-weighting, C-weighting, or flat response (the A-weighting is appropriate for lower noise levels, and the C-weighting for higher noise levels). Fast, slow, impulse, and peak response modes are available with selectable true RMS or peak detectors. Dual nonvolatile memories store or accumulate data for later recall. Internal nickel cadmium batteries power the IE-30A for about 3 hours continuously with a fast recharge cycle of 1.5 hours.

The system response, visible at a glance on the LED matrix, is useful in lining up sound reinforcement systems. A probe accessory which plugs into the microphone socket makes the instrument useful as a general test instrument. Noise surveys and real time analysis are obvious acoustical functions. There is a gated mode operation which comes into its own when the companion unit described below is attached. A review of the Ivie Model IE-30A may be found in Reference 139.

IVIE IE-17A

The Ivie 17-A microprocessor audio analyzer is an accessory to the Ivie IE-30A as shown in Fig. 17-4. Together they make an

322

impressive acoustical analyzing system that weighs about 5 pounds and can be held in one hand. For example, they can measure, calculate, store, display, average, and plot reverberation time in ⅓ or 1/1 octave bands over a range of 10 milliseconds to 99.9 seconds with 10 millisecond resolution. The IE-17A automatically calculates reverberation time and displays slope changes for small segments of the decay curve. A built in source control provides 1/3 or 1/1 octave bandwidth filters for prefiltering room signals.

Room delays between direct and reflected components can be measured accurately to tenths of milliseconds. The gating feature

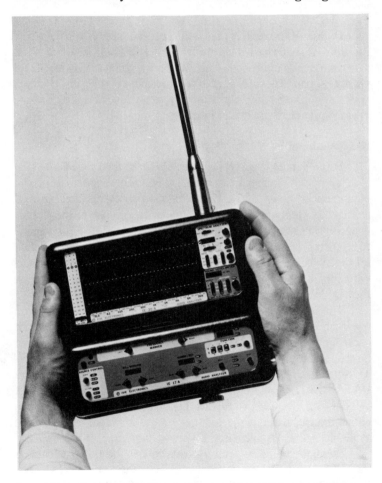

Fig. 17-4. The Ivie Model IE-30A (top) and the Model IE-17A (bottom) octave spectrum analyzer and microprocessor audio analyzer.

allows any given reflection to be isolated, measured, and compared to the direct wave. A frequency marker generated in the IE 17-A can be phase-locked to any channel of the IE-30A to allow high resolution measurements of 1/3 or 1/1 octave or wideband sound pressure level. This frequency marker can be stepped right or left to display the sound pressure level of any channel.

Tone burst testing includes control of zero crossing width of the burst, the precise time the IE-30A begins analysis, and the precise time the IE-30A stops analysis. Delay time up to 10 seconds (1 millisecond resolution) can be selected. Time delay spectrometry events can be analyzed and absorption coefficients measured as well as other time related phenomena.

Data on the screen of the IE-30A may be transferred to an X-Y recorder or strip chart recorder by a touch of a button. Reverberation time can be plotted in either of two ways, (1) calculated RT60 vs. the sound pressure level of the decaying signal in dB which reveals slope changes and room dynamic range, and (2) sound pressure level of the decay in dB vs. time.

Ivie Reverberation Measurements

Some details of the RT60 capabilities of the Ivie IE-30A/IE-17A combination are instructive. The interconnection of the equipment is shown in Fig. 17-5. The microphone may be removed from the IE-30A and placed on a tripod in the room to be measured. A power amplifier and loudspeaker are also required. If the space to be tested has an installed sound reinforcement system available for the measurements, the only additional equipment required is an extension cord and tripod and often the microphone can be left on the IE-30A.

The loudspeaker level is increased until the desired sound pressure level of the band limited noise is reached. A source control built into the IE-17A provides tracking 1/3 or 1/1 octave filters for prefiltering room signals. By pushing the TEST BUTTON on the IE-17A a reverberation decay is initiated and when the DATA READY light stops flashing the data appears in the DATA WINDOW. PROGRAM SELECTOR "A" controls are used to select the calculated data from the processor's memory files. The available values of calculated RT60 are shown in Table 17-1.

When the memory pointer points to 1-5, the RT60 calculated for the first 5 dB of decay is displayed in the DATA WINDOW. Pointing to 1-10 calls up the RT60 for the first 10 dB, 1-30 calls up the RT60 calculated for the first 30 dB. In other words, the

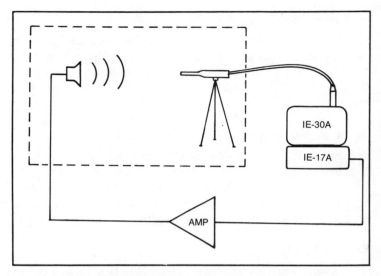

Fig. 17-5. Equipment arrangement for measuring room acoustics with the Ivie Model IE-30A and IE-17A analyzing system.

RT60 slope corresponding to any of the portions of the reverberation decay listed in Table 17-1 may be displayed in the DATA WINDOW. In this way decay slope changes may be called up and evaluated. If a reverberation time of 1.16 seconds is given for the second 5 dB of decay, this means that the slope corresponding to this segment of the decay *when extrapolated to 60 dB* gives 1.16 second.

Table 17-1. IE-30A/IE-17A Calculated RT60 Values.

Program Selector A	Calculated RT60 for:
1-5	The first 5 dB of decay
2-5	Second 5 dB of decay
3-5	Third 5 dB of decay
4-5	Fourth 5 dB of decay
5-5	Fifth dB of decay
6-5	Sixth 5 dB of decay
1-10	First 10 dB of decay
2-10	Second 10 dB of decay
3-10	Third 10 dB of decay
1-15	First 15 db of decay
2-15	Second 15 dB of decay
1-20	First 20 dB of decay
1-30	First 30 dB of decay

Plotting and Displaying RT60 Data

By connecting an X-Y recorder to the IE-17A, a printout of the reverberation decay stored in the memory may be had by pushing the PLOT BUTTON. The next push of this button causes the X-Y recorder to plot the calculated RT60 for each 5 dB segment along the decay curve.

Multiple RT60 Samples

Successive decays under identical conditions will not be identical because of the variability introduced by the randomness of the noise source. The Ivie IE-30A/17A instrument is capable of recording successive decays in the memory, automatically averaging all decays recorded. The average may then be plotted.

External Trigger

If an impulse sound source is used, the IE-30A/17A can be triggered by an external audio pulse. The smart IE-17A stores both rise and decay of sound in the room when triggered by such an external audio pulse. The computer searches for the maximum value and reconstructs only the decay portion of the curve and calculates the RT60 values in the usual way.

Gated Time Mode

The gated time mode in the IE-17A will enable the measurement of room and device response on an anechoic basis, but in non-anechoic rooms. By this technique absorption coefficients may be measured as well as the isolation of the direct component or any reflected component of a transmitted signal in room analysis. Not only that, but the changes of any of these parameters with frequency is possible.

The *pulse width* defines the length of time the signal source is gated on and is adjustable from 1 millisecond to 9.99 seconds in 1 millisecond steps. *Delay time* is that time from the leading edge of the triggered pulse to the instant the IE-30A begins analysis. The range is adjustable from 1.0 millisecond to 9.999 seconds in 1 millisecond steps. The *cycle time* is the time between signal source pulses and is adjustable from 0.00 second (for a single shot pulse) and from 0.01 to 30.00 seconds in 10 millisecond steps. *Aperture time* determines the length of time the IE-30A analysis continues. Its range is from 1.0 millisecond to 9.999 seconds in 1 millisecond steps.

With these four parameters under minute and precise control, any portion of a received signal (e.g., a reflection) can be selected by the IE-17A for analysis by the IE-30A. An oscilloscope is a very helpful accessory as it makes visible the relative relationship of the four parameters as well as the exact shape of the signal to be analyzed.

Swept Sine Measurements

With an X-Y recorder and a swept frequency oscillator the IE-30A/17A is able to perform measurements which might be called a slow version of time delay spectrometry. A sine wave source that is being swept very slowly (typically 1-3 minutes full range) is fed into the IE-17A gating system which processes it into a pulse which, in turn, is fed into the audio system under test. The IE-17A is then programmed to receive the direct component, a reflected component, or any other portion of the transmitted pulse. With each new pulse, the IE-17A performs a sample and hold on the sound pressure level of the signal and translates the measured amplitude to an exactly proportional dc voltage for the Y-axis of the plotter. The ramp output of the oscillator drives the X-axis after a calibration from the IE-17A.

Swept plots may be of the direct wave vs. frequency, or a plot of the direct wave compared to a reflection over the same frequency range. Just as feasible is a plot of the direct wave compared to the entire reverberant sound field vs. frequency. The variation of absorption coefficient with frequency and angle of incidence may be plotted. A review of the Ivie IE-30A audio analysis system may be found in Reference 139 and a review of the IE-17A in Reference 140.

SONY SOUND SYSTEM ANALYZER

The instruments described so far in this chapter have been designed primarily with the testing of electronic, recording, and sound system equipment in mind and, secondarily, acoustical measurements. This is quite understandable on the basis of economic justification and the fact that there is only a growing awareness of the importance of acoustical factors. At the convention of the Audio Engineering Society in Los Angeles in May, 1980, Sony engineers described a new apparatus, the Sound System Analyzer, for analyzing room acoustics in a way which has special appeal to the consultant in acoustics.[141] This instrument is pictured in Fig. 17-6. It measures only three things, but a strong case can be made that they

Fig. 17-6. The Sony sound system analyzer.

are the three most important parameters in judging the acoustical quality of a space. These three factors are (1) reverberation time, (2) transmission characteristics, and (3) definition, or clearness ratio (often called by the German name, "deutlichkeit").

Sound Source

Pulse measurements of rooms have been recognized as having great value for a long time. Firing blank cartridges (even small cannons in large halls) and bulky electrical spark discharge equipment have been used effectively, but, perhaps, not very efficiently. Exciting a space with steady state noise requires a bulky amplifier and heavy loudspeaker. The sound source used in the Sony Sound System Analyzer reflects growing interest in pulses such as the Hamming pulse, or tone-burst. A very short pulse has a wide spectrum. A steady state sine signal (a very long pulse, as it were) has an extremely narrow spectrum. The Hamming pulse is between these two extremes. By shaping a burst of about 6 cycles of any frequency in a very special way, as shown in Fig. 17-7, a spectrum of ⅓ octave width is obtained. The pulse is shaped by digital computer and in this way the ⅓ octave bandwidth is obtained without benefit of filters. Utilizing such pulses with a spaced duty cycle also reduces the amplifier power demand.

Description of Analyzer

The Sony Sound System Analyzer uses both analog and digital techniques. It measures:

328

—Reverberation time vs. frequency

—Sound transmission vs. frequency

—Definition vs. frequency

—Space averaging of the above from up to 128 different locations

—The reverberation decay curve in any frequency band.

A mini-printer built into the analyzer prints out tables and graphs. All measurements are processed in real time. The pulse source and power amplifier are contained in the analyzer. The only outboard accessories required are a loudspeaker and a microphone.

Reverberation

Reverberation decays fluctuate due to the beating of normal modes of different natural frequencies. The exact shape of a decay also depends upon the initial amplitudes and phase angles of the normal modes at the moment the room excitation is cut off. When a band of random noise is used to excite the room the initial amplitudes and phase angles are different for each decay. As pointed out in Chapter 8 the shape of successive decays (see Fig. 8-7) taken under identical conditions will be different because of the randomness of the exciting signal because nothing else has changed. This fluctuation of successive decays has, in the past, been overcome by laboriously averaging many decays.

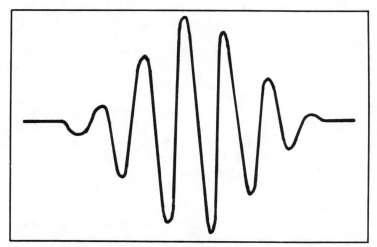

Fig. 17-7. The tone-burst shape utilized in the Sony sound system analyzer. This Hamming pulse has a spectrum width of 1/3 octave.

Schroeder[143] has described a new method of measuring reverberation time called the "integrated tone-burst method". With his method a single tone-burst decay yields a decay which is mathematically and rigorously identical to what one would get by averaging an infinite number of decays by the old method. The Sony analyzer uses Schroeder's method, hence a single decay at a given position is all that is necessary.

Sound Transmission

The sound transmission characteristics of a room are commonly measured by emitting pink noise from a loudspeaker, picking it up with a microphone located at some distance, and recording the response in 1/3 octave bands vs. frequency. The Sony Sound System Analyzer does this, in effect, by measuring the room's response to Hamming pulses of different frequencies. Comparisons of this method to the conventional method shows excellent agreement.

Definition

To understand the significance of "definition" we must go back to the Haas effect, remembering that all the reflected sound energy arriving at the ear during the 50 milliseconds immediately following the direct sound is integrated by the ear and perceived as strengthening and enhancing the direct sound. That which arrives later than 50 milliseconds is generally perceived as detrimental echoes. Meyer[144] coined the term "definition" and defined it as the ratio of the energy of the first 50 ms to the energy of the entire pulse:

$$D = \frac{E}{T} \times 100 \qquad (17\text{-}1)$$

where,

D = definition % or (clearness or deutlichkeit)
E = energy of pulse over the first 50 ms
T = total energy of pulse

A device to measure definition must be able (a) to square the sound pressure to get something proportional to energy, (b) to integrate energy over the first 50 ms of the pulse, (c) to integrate the total energy of the pulse, and (d) to divide (b) by (c). The Sony Sound System Analyzer does exactly this.

Operation

The loudspeaker and microphone are placed as desired. The input signal picked up by the microphone is fed to a peak detector which detects the maximum level of the input signal. This signal, after passing through an analog to digital converter, sets the gain of a programmable amplifier. This assures optimum level of the pulse for analysis.

Display

The mini-printer located on the top of the cabinet prints out the data in the form of either a table or a graph as selected by switch. The tabulation of sound transmission data is in the form of sound pressure level in decibels for each 1/3 octave from 50 Hz to 1 kHz. The definition is shown in percent for the same frequencies. Reverberation time in seconds is listed for each frequency.

An alternate form of the display is in the form of a bar graph, the length of the stack of zeros being the bar length. The transmission graph uses 2 dB for each zero in the stack. In the definition graph, each zero represents 5%. In the reverberation graph each zero represents 0.05 second. If these graph increments are considered too coarse, the more precise tables may be consulted. The space average of reverberation, transmission, and definition may be printed out by the device from up to 128 different locations. Further, the reverberation decay for any frequency band can be printed out on command.

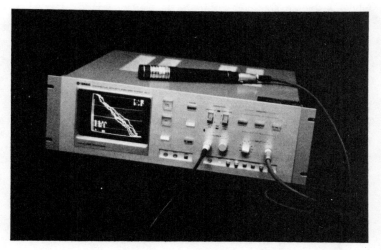

Fig. 17-8. The Yamaha AS-1 system for room acoustical measurements.

YAMAHA AS-1

At the same session of the Audio Engineering Society Convention (May, 1980) at which the Sony device was described Yamaha engineers presented their AS-1 system for room acoustical measurements (Yamaha name is associated with Nippon Gakki, Ltd.).[142] This very compact, portable, self-contained, and powerful piece of equipment utilizing the microprocessor and peripherals is illustrated in Fig. 17-8. It is designed to make field measurements in real time of reverberation time, transmission-frequency characteristics of a room, definition, and early decay time.

Configuration

The Yamaha AS-1 is made up of 5 basic bits of hardware, (1) the tone burst generator with its digital to analog converter and power amplifier, (2) microphone amplifier with its analog to digital converter, digital squarer, and integrator, (3) an 8 bit microprocessor unit with read only memory (ROM) and random access memory (RAM), (4) graphic display unit built around a 5-inch black and white cathode ray tube, and (5) control panel for user interface. The overall size of the basic box is only 5.5 x 15.7 x 11.8 inches.

Sound Source

The Sony device used a tone burst shaped after the Hamming plan. The Yamaha AS-1 also uses a tone burst but it is shaped after the Hanning plan. Even as the names are similar, so are the pulses whose energy is confined within the desired bandwidth. The AS-1 offers a choice of 1/3 octave or 1/1 octave bandwidths by the expedient of varying the number of cycles in the shaped pulse.

Squarer and Integrator

The philosophy of the designers of the AS-1 is that instead of sound pressure, sound pressure squared should be used for all measurements, even for reverberation time and as a real time analyzer. Integration is required for the definition measurement in which the energy in the first 50 milliseconds is compared to that of the entire pulse. The integration time is adjustable from 64 ms to 65.5 seconds.

Graphic Display

The cathode ray tube uses a raster scanning plan. Because of the tube's small face the resolution of the display has been reduced

332

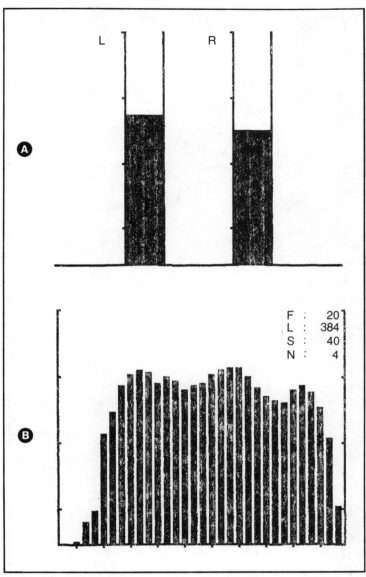

Fig. 17-9. Readout format of the Sony room acoustical analyzer, (A) for peak readings, (B) for room transmission test.

to 128 x 108 dots. This reduces the circuit size. The circuit has been further reduced by limiting the characters used in the display and by generating them all by software. If hard copies are required they may be obtained by the use of an optional, external, compact

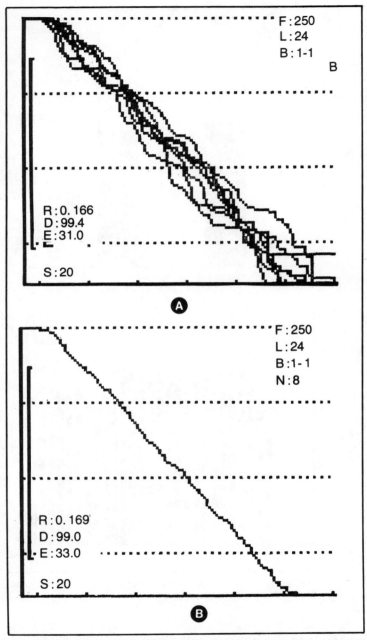

Fig. 17-10. Reverberation readout of the Yamaha room acoustics analyzer, (A) eight superimposed reverberation decays, and (B) the average of the eight decays of (A).

video plotter. They are considering incorporating a microminiature printer in the AS-1.

Control Panel

The switches on the panel are scanned by the microprocessor unit and under its control the switches can be used in common in different measurement modes. This reduces size and weight. All of the basic AS-1 circuits are on six 4″ x 8″ printed circuit boards.

Software

The above hardware is all under the control of software. In this way reverberation time, definition, and early decay time (for the first 10 dB of decay) are computed from the 128 bits of data sampled. The RT60 mode utilizes the principal part of the program. The software (1) sets up the parameters from panel switch positions, (2) accumulates background noise, (3) accumulates the square of the tone bursts, and (4) performs the arithmetic computations for the three functions mentioned above as well as the decay curves.

Measurements

After the power switch turns the AS-1 on, pushing the PK (peak meter) mode button brings the AS-1 display into the format shown in Fig. 17-9A. The TF (transfer function) mode button yields a room transmission display of the type shown in Fig. 17-9B with the pertinent data appearing in the alpha numerics. The RT (reverberation time) mode gives a display of the type shown in Fig. 17-10. The alpha numerics in this case translate as follows: F(frequency) 250 Hz, L(pulse length=(6 cycles)(1/250)=24 ms, B(bandwidth)=1/1 octave, N (number of decays averaged) = 8, R (reverberation time) = 0.169 sec, D (definition) = 99.0%, E (early decay time for first 10 dB of decay) = 33 ms, and S (time scale division value) = 20 ms. The ordinate is 10 dB between major divisions.

The 8 reverberation decays of Fig. 17-10A are shown averaged in B. The decays of A vary because they were taken at different microphone positions. At a given position successive decays retrace each other accurately because the Schroeder method is used, hence only one decay per position is necessary. The AS-1 measures reverberation time over the approximate range 0.1 to 51 seconds. The AS-1 is an impressive piece of equipment with efficient use of computer capability and built-in flexibility.

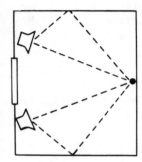

Appendix
Sound Absorption
Coefficients of General
Building Materials
and Furnishings

Complete tables of coefficients of the various materials[64] that normally constitute the interior finish of rooms may be found in the various books on architectural acoustics. Tables A1 and A2 give approximate values which will be useful in making simple calculations of the reverberation in rooms (courtesy of Ceilings & Interior Systems Contractors Association).

Table A-1. Absorption Coefficients.

Materials	125 Hz	250 Hz	500 Hz	1000 Hz	2000 Hz	4000 Hz
Brick, unglazed	.03	.03	.03	.04	.05	.07
Brick, unglazed, painted	.01	.01	.02	.02	.02	.03
Carpet,						
⅛" Pile Height	.05	.05	.10	.20	.30	.40
¼" Pile Height	.05	.10	.15	.30	.50	.55
3/16" combined Pile & Foam	.05	.10	.10	.30	.40	50
5/16" combined Pile & Foam	.05	.15	.30	.40	.50	.60
Concrete Block, painted	.10	.05	.06	.07	.09	.08
Fabrics						
Light velour, 10 oz. per sq. yd., hung straight, in contact with wall	.03	.04	.11	.17	.24	.35
Medium velour, 14 oz. per sq. yd., draped to half area	.07	.31	.49	.75	.70	.60
Heavy velour, 18 oz. per sq. yd. draped to half area	.14	.35	.55	.72	.70	.65
Floors						
Concrete or Terrazzo	.01	.01	.01	.02	.02	.02
Linoleum, asphalt, rubber or cork tile on concrete	.02	.03	.03	.03	.03	.02
Wood	.15	.11	.10	.07	.06	.07
Wood parquet in asphalt on concrete	.04	.04	.07	.06	.06	.07
Glass						
¼", sealed, large panes	.05	.03	.02	.02	.03	.02
24 oz., operable windows (in closed condition)	.10	.05	.04	.03	.03	.03
Gypsum Board, ½" nailed to 2x4's 16" o.c., painted	.10	.08	.05	.03	.03	.03
Marble or Glazed Tile	.01	.01	.01	.01	.02	.02
Plaster, gypsum or lime, rough finish or lath	.02	.03	.04	.05	.04	.03
Same, with smooth finish	.02	.02	.03	.04	.04	.03
Hardwood plywood paneling ¼" thick, Wood Frame	.58	.22	.07	.04	.03	.07
Water Surface, as in a swimming pool	.01	.01	.01	.01	.02	.03
Wood Roof Decking, tongue-and-groove cedar	.24	.19	.14	.08	.13	.10
Air, Sabins per 1000 cubic feet 50% RH				.9	2.3	7.2

Table A-2. Absorption of Seats and Audience.

	125 Hz	250 Hz	500 Hz	1000 Hz	2000 Hz	4000 Hz
Audience, seated, depending on spacing and upholstery of seats	2.5—4.0	3.5—5.0	4.0—5.5	4.5—6.5	5.0—7.0	4.5—7.0
Seats, heavily upholstered with fabric	1.5—3.5	3.5—4.5	4.0—5.0	4.0—5.5	3.5—5.5	3.5—4.5
Seats, heavily upholstered with leather, plastic, etc.	2.5—3.5	3.0—4.5	3.0—4.5	2.0—4.0	1.5—3.5	1.0—3.0
Seats, lightly upholstered with leather, plastic, etc.			1.5—2.0			
Seats, wood veneer, no upholstery	.15	.20	.25	.30	.50	.30
Wood pews, no cushions, per 18" length			.40			
Wood pews, cushioned, per 18" length			1.8—2.3			

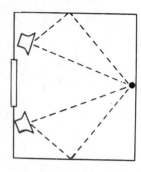

References

1. Wood, Alexander, *Acoustics*, Interscience Publishers, Inc.,
    New York (1941), Figure 11.4, page 302 (courtesy of
    Wiley-Interscience, Inc., New York).
2. Anon. *A Generator of Random Electrical Noise,* (A descrip-
    tion of Type 1390-B instrument) General Radio Com-
    pany Engineering Department Reprint No. E-110
    (1960).
3. Wiener, F.M., *On The Diffraction of a Progressive Wave By
    The Human Head*, Jour. Acoust. Soc. Am., Vol. 19, No.
    1 (1947), pp 143-146.
4. Wiener, F.M. and D.A. Ross, *Pressure Distribution in the
    Auditory Canal In A Progressive Sound Field*, Jour.
    Acoust. Soc. Am. Vol. 18, No. 2 (1946), pp 401-408.
5. Fletcher, Harvey, *Speech And Hearing in Communication*,
    D. Van Nostrand Company, Inc., New York (1953).
6. Van Bergeijk, Willem A., John R. Pierce, and Edward E.
    David, Jr. *Waves and The Ear*, Anchor Books, Double-
    day & Company, Inc., New York (1960).
7. Pierce, John R. and Edward E. David, Jr., *Man's World Of
    Sound*, Doubleday & Company, Inc., New York (1958).
8. Schroeder, M.R., D. Gottlob, and K.F. Siebrasse, *Com-
    parative Study of European Concert Halls: Correlation of
    Subjective Preference With Geometric And Acoustic
    Parameters*, Jour. Acoust. Soc. Am., Vol. 56, No. 4
    (October 1974), pp 1195-1201.
9. Hawkes, R.J. and Miss H. Douglas, *Subjective Experience In
    Concert Auditoria*, Acustica, Vol. 28, No. 5 (1971), pp
    235-250.

10. Meyer, Erwin, *Physical Measurements in Rooms And Their Meaning In Terms Of Hearing Conditions*, Proceedings of the Second International Congress On Acoustics, (1956), pp 59-68.

11. Silver, Sidney L., *The Psychoacoustic Aspects of Sound*, db Magazine, Vol. 7, No. 9 (September 1973), pp 33-37.

12. Fletcher, H. and W.A. Munson, *Loudness, Its Definition, Measurement and Calculation*, Jour. Acoust. Soc. Am., Vol. 5 (1933), pp 82-108.

13. Robinson, D.W. and R.S. Dadson, *A Re-Determination Of The Equal-Loudness Relations For Pure Tones*, British Jour. of App. Phys., Vol. 7 (1956), pp 166-181. Adopted by the International Standards Organization as Recommendation R-226.

14. Toole, Floyd E., *Loudness - Applications And Implications To Audio*, db Magazine, Part 1, Vol. 7, No. 5 (May 1973) pp 27-30; Part 2, Vol. 7, No. 6 (June 1973), pp 25-28.

15. Zwicker, G. Flottorp and S.S. Stevens, *Critical Bandwidth In Loudness Summation*, Jour. Acoust. Soc. Am., Vol. 29, No. 5 (May 1957), pp 548-557.

16. deBoer, E., *Note On The Critical Bandwidth*, Jour. Acoust. Soc. Am., Vol. 34, No. 7 (July 1962), p 985.

17. Anon. *Acoustics Handbook*, Application Note 100, Hewlett-Packard Company (1968).

18. Gardner, Mark B and Robert S. Gardner, *Problem Of Localization In The Median Plane: Effect Of Pinnae Cavity Occlusion*, Jour. Acoust. Soc. Am., Vol. 53. No. 2 (1973), pp 400-408.

19. Stevens, S.S. and J. Volkman, *The Relation of Pitch To Frequency: A Revised Scale*. Am Jour. Psychol., Vol. 53 (1940), pp 329-353.

20. Plomb, R. and H.J.M. Steeneken, *Place Dependence Of Timbre In Reverberant Sound Fields*, Acustica, Vol. 28, No. 1 (January 1973), pp 50-59.

21. Stark, Craig, *The Sense of Hearing*, Stereo Review (September), 1969, pp 66, 71-74.

22. Haas, Helmut, *The Influence Of A Single Echo On The Audibility of Speech*, Jour. Audio Engr. Soc., Vol. 20, No. 2 (March 1972), pp 146-159. This is an English translation from the German by Dr. Ingr. K.P.R. Ehrenberg of Haas' original paper in Acustica, Vol. 1, No. 2 (1951).

23. Fletcher, Harvey, *The Ear As A Measuring Instrument*, Jour. Audio. Engr. Soc., Vol. 17, No. 5 (October 1969), pp 532-534.

24. Anon. *The Relations of Hearing Loss To Noise Exposure*, A report by Exploratory Subcommittee Z-24-X-2 of the American Standards Association, Acoustical Society of America, sponsor.

25. Raichel, Daniel R., *Recreational Deafness - How Can Audio Engineers Stem It?*, Presented at the 64th convention of the Audio Engr. Soc. In New York City (1979). Preprint #1535 (I-6).

26. Anon., *General Industry: OSHA Safety And Health Standards (29CFR1910)* U.S. Department of Labor, Occupational Safety And Health Administration, OSHA 2206, revised November 7, 1978, U.S. Government Printing Office stock No. 029-015-00054-65.

27. Steinberg, John C. and Mark B. Gardner, *On The Auditory Significance Of The Term Hearing Loss*, Jour. Acoust. Soc. Am., Vol. 11. (January 1940), pp 270-277.

28. Knudsen, Vern O. and Cyril M. Harris, *Acoustical Designing in Architecture*, John Wiley & Sons, New York (1950).

29. Bartlett, Bruce, *A Scientific Explanation of Phasing (Flanging)*, Jour. Audio. Engr. Soc., Vol. 18, No. 6 (Dec 1970), p 674.

30. Burroughs, Lou, *Microphones: Design and Application*, Sagamore Publishing Co., Inc., Plainview, New York 11803 (1974). Chapters 10 and 11.

31. Davis, Chips and Don Davis, *(LEDE) Live End - Dead End Control Room Acoustics, (TDS) Time Delay Spectrometry, (PZM) Pressure Zone Microphones*, Recording Engineer/Producer, Vol. 10, No. 1 (Feb 1979), p 41. See also literature offered by Crown International, Inc. of Elkhart, Indiana 46514.

32. Everest, F. Alton, *The Complete Handbook of Public Address Sound Systems*, Tab Books, Blue Ridge Summit, PA 17214 (1978), pages 62-66.

33. Morse, Philip M. and Richard H. Bolt, *Sound Waves in Rooms*, Reviews of Modern Physics, Vol. 16, No. 2 (April 1944), pp 69-150.

34. Gilford, Christopher, *Acoustics For Radio and Television Studios*, Institution of Electrical Engineers, Monograph Series 11, Peter Peregrinus Ltd., London (1927), especially pages 120-125.

35. Gilford, C.L.S., *The Acoustic Design of Talks Studios and Listening Rooms,* Proc. IEE, Vol. 106, Part B, No. 27 (May 1959), pp 245-258. Reprinted in Jour. Audio Engr. Soc., Vol. 27, No. ½ (1979), pp 17-31.

36. Hunt, Frederick V., *Investigation of Room Acoustics by Steady-State Transmission Measurements,* Jour. Acous. Soc. Am., Vol. 10 (Jan 1939), pp 216-227.

37. Hunt, F.V., L.L. Beranek, and D.Y. Maa, *Analysis of Sound Decay in Rectangular Rooms,* Jour. Acous. Soc. Am., Vol. 11 (July 1939), pp 80-94.

38. Bolt, R.H., *Perturbation of Sound Waves in Irregular Rooms,* Jour. Acous. Soc. Am., Vol. 13 (July 1942), pp 65-73.

39. Bolt, R.H., *Note on Normal Frequency Statistics in Rectangular Rooms,* Jour. Acous. Soc. Am., Vol. 18, No. 1 (July 1946), pp 130-133.

40. Knudsen, Vern O., *Resonance in Small Rooms,* Jour. Acous. Soc. Am. (July 1932), pp 20-37.

41. Mayo, C. G., *Standing Wave Patterns in Studio Acoustics,* Acustica, Vol. 2, No. 2 (1952), pp 49-64.

42. Meyer, Erwin, *Physical Measurements in Rooms and Their Meaning in Terms of Hearing Conditions,* Proc. 2nd International Congress on Acoustics, (1956), pp 59-68.

43. Louden, M. M., *Dimension-Ratios of Rectangular Rooms With Good Distribution of Eigentones,* Acustica, Vol. 24 (1971), pp 101-104.

44. Sepmeyer, L. W., *Computed Frequency and Angular Distribution of the Normal Modes of Vibration in Rectangular Rooms,* Jour. Acous. Soc. Am., Vol. 37, No. 3, (March 1965), pp 413-423.

45. Bonello, Oscar John, *A New Computer Aided Method For The Complete Acoustical Design of Broadcasting and Recording Studios,* Trans. IEEE International Conference on Acoustics, Speech, and Signal Processing (1979), pp 326-329.

46. Bonello, Oscar John, *A New Criterion For the Distribution of Normal Room Modes,* Presented at the 64th Convention of the Audio Engr. Soc. (Nov 1979), Preprint #1530 (F-6).

47. Dubout, P., *Perception of Artificial Echoes of Medium Delay,* Acustica, Vol. 8 (1958), pp 371-378.

48. Haas, Helmut, *The Influence of a Single Echo on the Audibility of Speech,* Jour. Audio Engr. Soc., Vol. 20 No. 2 (March

1972), pp 146-159. This is an English translation from the German by Dr. Ing. K.P.R. Ehrenberg of Haas' original paper in Acustica, Vol. 1, No. 2 (1951).

49. Nickson, A.F.B., R.W. Muncey and P. Dubout, *The Acceptability of Artificial Echoes With Reverberant Speech and Music*, Acustica, Vol. 4 (1954), pp 515-518.

50. Matsudaira, T. Ken, et al, *Fast Room Acoustic Analyzer (FRA) Using Public Telephone Line and Computer*, Jour. Audio Engr. Soc., Vol. 25, No. 3 (March 1977), pp 82-94.

51. Peutz, V.M.A., *Articulation Loss of Consonants as a Criterion for Speech Transmission in a Room*, Jour. Audio Engr. Soc., Vol. 19, No. 11 (Dec. 1971), pp 915-919.

52. Klein, W., *Articulation Loss of Consonants as a Basis for the Design and Judgement of Sound Reinforcement Systems*, Jour. Audio Engr. Soc., Vol. 19, No. 11 (Dec. 1971), pp 920-922.

53. Beranek, L. L., *Music, Acoustics, and Architecture*, John Wiley & Sons, New York (1962).

54. Balachandran, C.G., *Pitch Change During Reverberant Decay*, Jour. Sound & Vibration, Vol. 48, No. 4 (1976), pp 559-560.

55. Rettinger, Michael, *Reverberation Chambers for Broadcasting and Recording Studios*, Jour. Audio Engr. Soc., Vol. 5, No. 1 (1957), pp 18-22. See also his *Note on Reverberation Chambers*, JAES, Vol. 5, No. 2 (1957), pg 108.

56. Putnam, Scott and Tom Lubin, *Construction of a Live Echo Chamber*, Recording engr./Prod., Vol. 10, No. 4, (Aug 1979), pp 73-74, 76, 78, 80, 81.

57. Eyring, C. F., *Reverberation Time in "Dead" Rooms*, Jour. Acous. Soc. Am., Vol. 1 (1930), pp 217-241.

58. Norris, R. F., *A Derivation of the Reverberation Formula*, Appendix II of *Architectural Acoustics* by V.O. Knudsen, John Wiley & Sons, N.Y. (1932).

59. Young, Robert W., *Sabine Reverberation Equation and Sound Power Calculations*, Jour. Acous. Soc. Am., Vol. 31, No. 7 (July 1959), pp 912-921. See also a letter to the editor Vol. 31, No. 12 (December 1959), p 1681.

60. Flanagan, J. L., *Voices of Men and Machines*, Jour. Acous. Soc. Am., Vol. 51, No. 5 (Part 1) (May 1972), pp 1375-1387.

61. Hutchins, Carleen M. and Francis L. Fielding, *Acoustical Measurement of Violins*, Physics Today (July 1968), pp 35-41. Contains extensive bibliography.

62. Sivian, L. J., H.K. Dunn, and S.D. White, *Absolute Amplitudes and Spectra of Certain Musical Instruments and Orchestras*, Jour. Acous. Soc. Am., Vol. 2, No. 3 (Jan 1931), pp 330-371.

63. Harris, Cyril M., *Acoustical Properties of Carpet*, Jour. Acous. Soc. Am., Vol. 27 No. 6 (Nov 1955), pp 1077-1082.

64. Anon., *Acoustical ceilings: Use and Practice*, Published (1978) by Ceiling & Interior Systems Contractors Association, 1800 Pickwick Avenue, Glenview, IL 60025. See appendix.

65. Anon., *Quiet Zone*, a brochure describing Sonex acoustical foam published by Charles Industries Corporation, Minneapolis, MN 55428.

66. Kingsbury, H. F. and W. J. Wallace, *Acoustic Absorption Characteristics of People*, Sound and Vibration, Vol. 2 No. 2 (Dec 1968), pp 15, 16.

67. Everest, F. Alton, *Acoustic Techniques for Home and Studio*, Tab Books, Blue Ridge Summit, PA 17214 (1973) pp 71-75.

68. Beranek, Leo L., *Acoustics*, McGraw-Hill Book Company, New York, N.Y. (1954), p/300.

69. Mankovsky, V.S., *Acoustics of Studios and Auditoria*, Communications Arts Books, Hastings House, New York, N.Y. 10016 (1971).

70. Siekman, William, private communication. Mr. Siekman was manager of Riverbank Acoustical Laboratories at that time and made the measurements for presentation as a paper before the Acoustical Society of America during its April, 1969, meeting.

71. Beranek, Leo. L., *Broadcast Studio Design*, Jour. of the Soc. of Mot. Pic. and TV Engrs., Vol. 64 (Oct. 1955), pp 550-559.

72. Rettinger, Michael, *Acoustic Design and Noise Control*, Chemical Publishing Company, Inc., New York, N.Y. (1973).

73. Rettinger, Michael, *Low Frequency Sound Absorbers*, db Magazine, Vol. 4, No 4 (April, 1970), pp 44-46.

74. Rettinger, Michael, *Low Frequency Slot Absorbers*, db Magazine, Vol. 10 No. 6 (June, 1976), pp 40-43.

75. Rettinger, Michael, *On The Acoustics of Multitrack Recording Studios*, J. Audio Engr. Soc., Vol. 19, No. 8 (September, 1971), pp 651-655.

76. Callaway, D. B., and L. G. Ramer, *The Use of Perforated Facings in Designing Low Frequency Resonant Absorbers*, Jour. Acous. Soc. Am., Vol. 24, No. 3, (May, 1952), pp 309-312.

77. Veale, Edward J., *The Environmental Design of a Studio Control Room*, Preprint No. A-2(R), 44th Audio Engineering Society Convention, 20-22 (February 1973), Rotterdam.

78. Van Leeuwen, F. J., *The Damping of Eigentones in Small Rooms By Helmholtz Resonators*, European Broadcast Union Review, A. 62 (1960), pp 155-161.

79. Gilford, Christopher, *Acoustics for Radio and Television Studios*, IEE Monograph Series 11, Peter Peregrinus, Ltd., London (1972), pp 149-157.

80. Randall, K.E. and F.L. Ward, *Diffusion of Sound in Small Rooms*, Proc. Inst. of Elect. Engrs., Vol. 107B (Sept 1960), pp 439-450.

81. Bolt, R.H., *Note on Normal Frequency Statistics For Rectangular Rooms*, Jour. Acous. Soc. Am., Vol. 18, No. 1 (July 1946), pp 130-133.

82. Volkmann, J.E., *Polycylindrical Diffusers in Room Acoustical Design*, Jour. Acous. Soc. Am., Vol. 13, (1942), pp 234-243.

83. Boner, C. P., *Performance of Broadcast Studios Designed With Convex Surfaces of Plywood*, Jour. Acous. Soc. Am., Vol. 13 (1942), pp 244-247.

84. Sepmeyer, L. W., *Computed Frequency and Angular Distribution of The Normal Modes of Vibration in Rectangular Rooms*, Jour. Acous. Soc. Am., Vol. 37, No. 3 (March 1965), pp 413-423.

85. Louden, M. M., *Dimension-Ratios of Rectangular Rooms With Good Distribution of Eigentones*, Acustica, Vol. 24 (1971), pp 101-103.

86. Nimura, Tadamoto and Kimio Shibayama, *Effect of Splayed Walls of a Room on the Steady-State Sound Transmission Characteristics*, Jour. Acous. Soc. Am., Vol. 29, No 1, (Jan 1957), pp 85-93.

87. Somerville, T. and F.L. Ward, *Investigation of Sound Diffusion in Rooms by Means of a Model*, Acustica, Vol. 1, No 1, (1951), pp 40-48.

88. Schroeder, M. R., *Diffuse Sound Reflection by Maximum-Length Sequences,* Jour. Acous. Soc. Am., Vol. 57, No. 1 (Jan 1975), pp 149-150.

89. Schroeder, M.R. and R.E. Gerlach, *Diffuse Sound Reflection Surfaces*, 9th International Congress on Acoustics, Madrid (1977), Paper D8.

90. Schroeder, M.R., *Binaural Dissimilarity and Optimum Ceilings for Concert Halls: More Lateral Sound Diffusion,* Jour. Acous. Soc. Am., Vol. 65, No. 4, (Apr 1979), pp 958-963.

91. Beranek, L. L., *Revised Criteria For Noise in Buildings,* Noise Control, Vol. 3., No. 1 (January 1957), pp 19-27. The originally proposed curves have been slightly changed to incorporate the American Standard preferred frequencies and newer data on the threshold of hearing. See Schultz, T.J., Jour. Acous. Soc. Am., Vol. 43, No. 3 (1968), pp 637-8.

92. Anon., *ASHRAE Handbook and Product Directory - 1976 - SYSTEMS,* published by the American Society of Heating, Refrigerating, and Air-Conditioning Engineers, Inc., 334 East 47th Street, New York, NY 10017. Chapter 35 - *Sound and Vibration Control.*

93. Anon., *ASHRAE Handbook and Product Directory - 1977 - FUNDAMENTALS,* Chapter 7, *Sound Control Fundamentals.*

94. Doelling, Norman, *How Effective Are Packaged Attenuators?* ASHRAE Journal, Vol. 2, No. 2 (Feb 1960), pp 46-50.

95. Sanders, Guy J., *Silencers: Their Design and Application,* Sound and Vibration, Vol. 2, No. 2 (Feb 1968), pp 6-13.

96. Jackson, G.M. and H.G. Leventhall, *The Acoustics of Domestic Rooms,* Applied Acoustics, Vol. 5 (1972), pp 265-277.

97. Kishinaga, Shinji, Yashushi Shimizu, Shigeo Ando, and Kiminori Yamaguchi, *On The Room Acoustic Design Of Listening Rooms.* Presented at the 64th convention of the Audio Engr. Soc. (Nov 1979), Preprint No. 1524 (F-7).

98. Cook, Emory, *Evaluating The Influence of Room Acoustics on L/R Stereo Perception,* Presented at the 61st convention of the Audio Engr. Soc. (Nov 1978), Preprint No. 1385 (G-4).

99. _____ Such as Pink Noise Test Record Type QR 2011 offered by Bruel & Kjaer Instruments, Inc., whose main office is located at 185 Forest Street, Marlborough, MA 01752, with field offices in many metropolitan centers.

100. Snow, William B., *Application of Acoustical Engineering Principles To Home Music Rooms*, Inst. of Radio Engrs. Transactions on Audio, Vol. AU-5, No. 6 (Nov-Dec 1957), pp 153-159.

101. Allison, Roy F. and Robert Berkovitz, *The Sound Field in Home Listening Rooms*, Jour. Audio Engr. Soc., Vol. 20, No 6 (July/Aug 1972), pp 459-469.

102. Kuhl, Walter, *Optimal Acoustical Design of Rooms For Performing, Listening, and Recording*, 2nd Internat. Congress on Acous., (1956), pp 53-58.

103. Everest, F. Alton, *Acoustic Techniques For Home and Studio* (1973), TAB Books #646, Blue Ridge Summit, Pa 17214.

104. Bruce, Robert H., *How To Construct Your Own Studio In One Easy Lesson*, Presented at the 57th Convention of the Audio Engr. Soc. (May 1977), Preprint No. 1245 (J-3).

105. Dilley, Michael S., *Producer's Studio: A Do-It-Yourself Construction Project*, db The Sound Engr. Magazine, Vol. 13, No. 7 (July 1979), pp 26-35.

106. Penner, PS., *Acoustic Specification and Design of Supreme Being Studios*, Jour. Audio Engr. Soc, Vol. 27, No. 5 (May 1979), pp 351-367. Includes 134 references to the literature.

107. Storyk, John and Robert Wolsch, *Solutions To 3 Commonly Encountered Architectural and Acoustic Studio Design Problems*, Recording Engr/Prod, Vol. 7, No. 1 (Feb 1976), p 11.

108. Rettinger, Michael, *Recording Studio Acoustics*, db The Sound Engr. Magazine:
    Part 1: Vol. 8, No. 8 (Aug 1974) pp 34-47
    Part 2: Vol. 8, No. 10 (Oct 1974) pp 38-41
    Part 3: Vol. 8, No. 12 (Dec 1974) pp 31-33
    Part 4: Vol. 9, No. 2 (Feb 1975) pp 34-36
    Part 5: Vol. 9, No. 4 (Apr 1975) pp 40-42
    Part 6: Vol. 9, No. 6 (June 1975) pp 42-44.

109. Everest, F. Alton, *How To Build a Small Budget Recording Studio From Scratch*, (1979) TAB Books #1166, Blue Ridge Summit, PA 17214.

110. Woram, John, *Anyone for Two-Track?* db The Sound Eng. Magazine, Vol.4, No. 2 (Feb 1970), pp 16-17.

111. Alexandrovich, George, *Multi-Channel Recording - Why?* db The Sound Engr. Magazine, Vol. 3, No. 3 (March 1969), pp 4,6,8 and Vol. 3 No. 4 (Apr 1969), pp 4,6.

112. Rettinger, Michael, *On The Acoustics of Multitrack Recording Studios*, Jour. Audio. Engr. Soc., Vol. 19, No. 8 (Sept 1971), pp 651-655.

113. Runstein, Robert E., *Modern Recording Techniques*, (1974), Howard W. Sams/Bobbs-Merrill, Indianapolis, Indiana.

114. Woram, John, *The Recording Studio Handbook*, Sagamore Publishing Co., Plainview, NY 11803 (1976)

115. Eargle, John M., *Sound Recording* (1976), Van Nostrand Reinhold., New York, NY.

116. Everest. F. Alton, *Handbook of Multichannel Recording*, (1975), TAB Book No. #781, Blue Ridge Summit, PA 17214.

117. Queen, Daniel, *Monitoring Room Acoustics*, db The Sound Engr. Magazine, Vol. 7, No. 5 (May 1973) pp 24-26.

118. Rettinger, Michael, *On The Acoustics of Control Rooms*, Presented at the 57th convention of the Audio Engr. Soc. (May 1977), Preprint No. 1261 (J-1).

119. Davis, Don and Carolyn Davis, *Putting It All Together In a Control Room*, Synergetic Audio Concepts, Techtopics, Vol. 5, No. 7 (April 1978), pp 1-4.

120. Davis, Don and Chips Davis, *The "LEDE" Concept For The Control of Acoustic and Psychoacoustic Parameters In Recording Control Rooms*, Presented at the 63rd convention of the Audio Engr. Soc. (May 1979), Preprint No. 1502 (F/2).

121. Richards, Randy and Gaston Nichols, *Translating LEDE Control Room Design Into Practical Experience*, Presented at the 66th convention of the Audio Engr. Soc. (May 1980), Preprint No. 1631 (G-3).

122. Davis, Don, *The Role of The Initial Time Delay Gap in The Acoustic Design of Control Rooms For Recording and Reinforcing Systems*, Presented at the 64th convention of the Audio Engr. Soc. (Nov 1979), Preprint No. 1547 (F-3).

123. Rettinger, Michael, *A Live-End Environment For Control Room Loudspeakers*, db The Sound Engr. Magazine, Vol. 14, No. 6 (June 1980), pp 42-43. See also Rettinger's letter to the editor, Recording Engr./Prod., Vol. 11, No. 2 (April 1980), pp 14,16.

124. Rettinger, Michael, *Control Room Acoustics,* db The Sound Engr. Mag., Vol. 11, No. 4 (April 1977), pp 26-29.

125. Fierstein, Alan, *Optimizing Control Room Reverberation,* Recording Engr./Prod., Vol. 10, No. 4 (Aug 1979), pp 87-91.

126. Long, E.M., *A Time-Align Technique For Loudspeaker System Design,* Presented, the 54th convention of the Audio Engr. Soc. (May 1976), Preprint No. 1131 (M-8).

127. Eargle, John, *Equalizing The Monitoring Environment,* Jour. Audio Engr. Soc., Vol. 21, No. 2 (March 1973), pp 103-107.

128. Eargle, John M., *A Survey of Recording Studio Monitoring Problems,* Presented at the 44th convention of the Audio Engr. Soc. (Feb 1973), Preprint No. A-3(R).

129. Eargle, John, *Requirements For Studio Monitoring,* db The Sound Engr. Magazine, Vol. 13, No. 2 (Feb 1979) pp 34-37.

130. Smith, Pete, *Room Tuning,* Studio Sound, Vol. 21, No. 6, (June 1979), pp 58-62.

131. Fierstein, Alan, *The Equalization Myth,* Recording Engr./ Prod. Vol. 8 No. 3 (June 1977) pp 47-49. Also in db The Sound Engr. Magazine, Vol. 11, No. 8 (Aug 1977), pp 32-33.

132. Borenius, Juhani and Seppo V. Korhonen, *Standardized Listening Conditions In Sound Control Rooms,* Presented at the 59th convention of the Audio Engr. Soc. (Feb 1978), Preprint No. 1332 (G-8).

133. ————, The Pressure Recording Process (PRP) (patent pending) is the basis for the Pressure Zone Microphone (PZM) originally manufactured by Wahrenbrock Sound Associates, Ltd. Crown PZ Microphones are now manufactured under license from E.M. Long Associates and their agent Synergetic-Audio-Concepts. PZMicrophones, PZM, and Pressure Zone Microphones are trademarks of Crown International, Inc., 1718 W. Mishawaka Road, Elkhart, Indiana 46514.

134. Andrews, David M., *Pressure Zone Microphones™, A Practical Application of the Pressure Zone Recording Process™,* Presented at the 66th Convention of the Audio Engr. Soc. (May 1980), Preprint No. 1647 (J-4).

135. Heyser, R.C., *Acoustical Measurements By Time-Delay Spectrometry,* Jour. Audio Engr. Soc., Vol. 15, No. 4 (Oct 1967), pp 370-382.

136. Sommerwerck, William, *An Audio Micro-Computer For Tests And Measurements,* db The Sound Engineering Magazine, Vol. 14, No. 2 (Feb 1980), pp 28-32.

137. Ford, Hugh *Review: Inovonics Model 500 Acoustic Analyser,* Studio Sound, Vol. 21, No. 4 (April 1979), pp 88-92.

138. Smith, Peter, *Operational Assessment: Inovonics Model 500 Acoustic Analyser,* Studio Sound, Vol. 21, No. 5, (May 1979), p 90.

139. Ford, Hugh, *Review: Ivie IE-30A Audio Analysis System,* Studio Sound, Vol. 21, No. 11 (Nov 1979), pp 78-82.

140. Ford, Hugh, *Review: Ivie IE-17A Microprocessor Audio Analyser,* Studio Sound, Vol. 21, No. 11 (Nov 1979), pp 84-88.

141. Takise, T., T. K. Matsudaira, and H. Nakajima, *A New Apparatus To Analyze The Acoustics of a Room,* Presented at the 66th Convention of the Audio Engr. Soc. (May 1980), Preprint No 1619 (E-2).

142. Kawakami, Fukushi, Koji Niimi, and Kiminori Yamaguchi, *A Tool For Room Acoustic Measurement,* Presented at the 66th Convention of the Audio Engr. Soc. (May 1980), Preprint No 1632 (E-1).

143. Schroeder, M.R., *New Method of Measuring Reverberation Time,* Jour. Acous. Soc. Am., Vol. 37, No. 37 (1965), pp 409-412.

144. Meyer, E., *Definition and Diffusion in Rooms,* Jour. Acous. Soc. AM., Vol 26, No 5 (Sept 1954), pp 630-636.

145. Bucklein, R, *The Audibility of Frequency Response Irregularities,* Jour. Audio Engr. Soc., Vol. 29, No. 3 (March 1981), pp 126-131.

146. Rodgers, C.A. Puddie,*Pinna Transformations and Sound Reproduction,* Jour. Audio Engr. Soc., Vol. 29, No. 4 (April 1981), pp 226-234. Rodgers hypothesizes that comb filtering between the direct sound and reflections from the convolutions of the outer ear (pinna) produces the cues for direction of arrival of the sound.

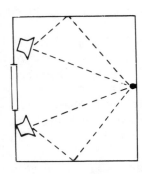

Index

Edited by Roland Phelps